Evolution of the Genetic Code

Evolution of the Genetic Code

SYOZO OSAWA

Executive Researcher, Biohistory Research Hall and
Emeritus Professor of Molecular Genetics, Nagoya University, Japan

This publication was supported by a generous donation from the Daido Life Foundation.

Oxford New York Tokyo
OXFORD UNIVERSITY PRESS

Oxford University Press, Walton Street, Oxford OX2 6DP
Oxford New York
Athens Auckland Bangkok Bombay
Calcutta Cape Town Dar es Salaam Delhi
Florence Hong Kong Istanbul Karachi
Kuala Lumpur Madras Madrid Melbourne
Mexico City Nairobi Paris Singapore
Taipei Tokyo Toronto
and associated companies in
Berlin Ibadan

Oxford is a trade mark of Oxford University Press

Published in the United States
by Oxford University Press Inc., New York

First published 1995
Reprinted 1995 (with corrections)

A catalogue record for this book is available from the British Library

Library of Congress Cataloging in Publication Data
Ōsawa, Syōzō, 1928–
Evolution of the genetic code / Syozo Osawa.
Includes bibliographical references and index.
1. Evolutionary genetics. I. Title.
QH390.083 1995 575–dc20 94-45539
ISBN 0 19 854781 1

Printed in Great Britain by
Bookcraft (Bath) Ltd
Midsomer Norton, Avon.

Preface

The genetic code is essential to all forms of life and is of fundamental importance to the whole of biology. Until quite recently, the code was thought to be invariable, 'frozen' in all organisms, because of the way in which any change would produce widespread alterations in the amino acid sequences of proteins. The universality of the genetic code was first challenged in 1981, when mammalian mitochondria were found to use a code which deviated somewhat from the universal. It was thought that either the mammalian mitochondrial code represented a remnant of an ancient code, or that the change in the code happened to be tolerable in mitochondria because of their small genome (ten or so genes).

In 1984, my research group at Nagoya University found that a bacterium, *Mycoplasma capricolum*, used a deviant genetic code, namely that UGA, a universal stop codon, was read as tryptophan. This finding was communicated by Dr Motoo Kimura of the National Institute of Genetics, Japan, in the 1985 January issue of the *Proceedings of the Japan Academy*, and in the 1985 April issue of *Proceedings of the National Academy of Sciences, USA*.

Surprisingly, at about the same time, French, American, and German workers announced that some ciliated protozoans used UAA and UAG as glutamine codons.

These findings, together with the deviant genetic codes in mitochondria, showed that the genetic code, formerly thought to be frozen, is, in fact, in a state of evolution.

Obviously, a new theory was needed to account for the changes in codon meanings. Accordingly, Dr Thomas H. Jukes of the University of California, Berkeley, and I proposed the codon capture theory in 1989. The theory was based on experimental and theoretical studies conducted by us, in addition to data available at that time. In this book, a particular emphasis has been placed upon a series of studies on the evolution of the code that were performed in our laboratory between 1982 and 1992.

I would express my wholehearted appreciation to all my outstanding co-workers at Nagoya University, and to Dr Thomas H. Jukes, who has been collaborating with me for many years.

In the preparation of this book valuable suggestions and help have been given by Dr Dolf Hatfield, Dr Thomas H. Jukes, Dr Kimitsuna Watanabe, Dr Shigeyuki Yokoyama, and Dr Takeshi Ohama. I am most grateful for their kindness. Cordial thanks are also due to those friends who generously offered me the fine photographs that have been used in the plates, and to colleagues for allowing me to reproduce figures or tables from their original papers. Their names are given in the legend of the appropriate figure or table.

This book was written between 1993 and 1994 at the Biohistory Research Hall, Takatsuki, Osaka. I wish to thank Dr Tokindo Okada, the director of the Hall, and other members there for their hospitality. My special thanks are due to Mrs Yayoi Takahashi for her devoted editorial work. Without her help, this book could not have been completed. Finally, I should mention that this monograph was conceived when Oxford University Press asked me to write a review of my work on the evolution of the genetic code on the occasion of my retirement from Nagoya University. I greatly appreciate all the expert advice and assistance I received from the publisher while I was writing this book.

Takatsuki S.O.
June 1994

To Tom Jukes

Contents

List of plates xiii

List of abbreviations xiv

1. The genetic code—history 1

1.1 Introduction 1
1.2 The day before yesterday 1
1.2.1 Template hypothesis in early days 1
1.2.2 The overlapping code 2
1.2.3 The commaless code 3
1.2.4 The adaptor hypothesis 5
1.2.5 tRNA as an adaptor 5
1.2.6 Triplet nature and fixed reading frame of the code 6
1.2.7 The discovery of messenger RNA 7
1.3 Yesterday 7
1.3.1 The breakthrough 7
1.3.2 Deciphering the amino acid code 7
1.3.3 Termination codons 8
1.3.4 Initiation codons 9
1.3.5 The genetic code *in vivo* 9
1.3.6 The wobble hypothesis 11
1.3.7 tRNA 11

2. The structure of the 'universal' genetic code 14

2.1 Codons 14
2.2 Codon–anticodon pairing rules 15
2.2.1 G 17
2.2.2 U 17
2.2.3 I 21
2.2.4 A 22
2.2.5 C 24
2.3 Stop codons and release factors 25
2.4 Four-way wobbling versus the 'two-out-of-three' mechanism 25
2.5 Role of nucleotide 37 in codon–anticodon pairing 26
2.5.1 U-started codons 26
2.5.2 A-started codons 27
2.5.3 C-started codons 27
2.5.4 G-started codons 31

2.6 tRNA identity 31
2.6.1 Organisms 31
2.6.2 Mitochondria 34

3. Anticodon composition 36

3.1 Bacteria 36
3.2 Chloroplasts and mitochondria 37
3.3 Resemblance between *Mycoplasma* spp. and mitochondria 37
3.4 Eukaryotes 40

4. Codon usage 45

4.1 Genomic G + C contents and codon usage 45
4.1.1 Directional mutation pressure 45
4.1.2 Bacteria 46
4.1.3 Eukaryotes 51
4.1.4 Mitochondria 52
4.2 Codon selection by tRNA 53
4.3 Stop codon selection by release factors 56

5. Unassigned or nonsense codons 58

5.1 Genesis of unassigned codons from amino acid codons 58
5.2 Unassigned codons and stop codons 63

6. The evolving genetic code 71

6.1 The frozen-accident theory 71
6.2 After the frozen-accident theory 71
6.3 Distribution of non-universal codons and their translation 74
6.3.1 The nuclear code 74
6.3.2 The mitochondrial code 82
UGA for Trp 86
AUA for Met 88
AAA for Asn 88
AGR for Ser, Gly, or stop 89
CUN for Thr 91
UAA for Tyr 92
6.4 Mechanisms of code change 93
6.4.1 Codon capture theory 93
6.4.2 Codon capture 94
6.5 Nuclear systems 96
6.5.1 UGA from stop to Trp 96
6.5.2 UAR from stop to Gln 99
6.5.3 UGA from stop to Cys 99
6.5.4 CUG from Leu to Ser and as a reversal to Leu 99

6.6 Mitochondrial systems 103
6.6.1 UGA from stop to Trp 103
6.6.2 CUN from Leu to Thr 103
6.6.3 AUA from Ile to Met 106
Yeast mitochondria 107
Metazoan mitochondria 107
6.6.4 AGR from Arg to Ser, Gly or stop 108
6.6.5 AAA from Lys to Asn; AUA from Met to Ile as a reversal of Ile to Met; UAA from stop to Tyr 113

7. Selenocysteine is coded by UGA 116

7.1 UGA as a selenocysteine codon 116
7.2 Pathway of selenocysteine biosynthesis 118
7.3 Structure of selenocysteine tRNA 119
7.4 Translational context of selenocysteine UGA 121
7.5 Evolution of UGA-coded selenocysteine synthesizing system 124

8. RNA editing 126

8.1 Biological meaning of RNA editing 126
8.2 Green plant mitochondria 126
8.2.1 Is CGG for Trp in plant mitochondria? 126
8.2.2 A model for the evolution of RNA editing 127
The appearance of RNA editing 127
Mutations at editable positions 129
Maintenance of RNA editing activity 129
8.3 Chloroplasts 131
8.4 Eukaryotic nuclear systems 132
8.5 Mitochondria 132
8.6 Insertion/deletion editing 132

9. Origin and early evolution of the genetic code 133

9.1 So many theories 133
9.1.1 Abiotic synthesis of amino acids 133
9.1.2 Two different aspects of the evolution of the genetic code 135
9.1.3 Stereochemical explanation of the origin of the genetic code 136
9.1.4 RRY and RNY hypotheses 137
9.1.5 RNA code 139
9.1.6 Ribozyme origin of the genetic code 141
9.1.7 Some other theories 145
9.1.8 Evolution of tRNA and amino acyl tRNA synthetase 147
9.1.9 Evolution of the ancient genetic code 147
9.1.10 Summary 152

9.2 The early genetic code 153
9.3 From the early genetic code to the universal genetic code 155
9.3.1 Problems to be solved 155
9.3.2 Reduction in the numbers of Trp and Met codons 156
9.3.3 Introduction of the UGA stop codon 156
9.3.4 Reassignment of codon AUA for Ile 157
9.3.5 Evolutionary diversification of anticodons 157
9.3.6 Origin of inosine-containing anticodons 158

10. Amino acid composition of proteins and the genetic code 161

10.1 Amino acid composition of proteins—the role of the genetic code 161
10.2 Selection against the genetic code—functional constraint 162
10.3 Selection against the genetic code—constraint by directional mutation pressure 165
10.4 Amino acid usage in proteins affects the levels of tRNAs 168

Epilogue 171

References 178

Index 199

Plates

The plates fall between pages 82 and 83.

Plate 1	A plaque showing the creation of the animals (southern Italian, *c.* 1080)
Plate 2	Rat liver mitochondria
Plate 3	*Mycoplasma capricolum* (colonies)
Plate 4	*Spiroplasma kunkelii* var. *allistephi*
Plate 5	*Candida parapsilosis*
Plate 6	*Candida zeylanoides*
Plate 7	*Candida cylindracea*
Plate 8	*Candida rugosa*
Plate 9	*Candida melibiosica*
Plate 10a and b	*Paramecium caudatum*
Plate 11	*Tetrahymena thermophilia*
Plate 12	*Stylonichia lemnae*
Plate 13	*Oxytricha* sp.
Plate 14	*Euplotes woodrufii*
Plate 15	*Acetabularia* sp.

Abbreviations

For the three-letter abbreviations of amino acids, see Table 1.1, p. 10.

N	A, G, U(T), or C
R	A, or G
Y	U(T), or C
f^5C	5-formylcytidine
I	inosine (base hypoxanthine)
L	lysidine (2-lysylcytidine)
^+U	xo^5U (see Table 2.1, p. 16, and Fig. 2.2, p. 19)
*U	xm^5s^2U, $cmnm^5U$, mcm^5U, or Um (see Table 2.1, p. 16, and Fig. 2.2, p. 19)
ARS	aminoacyl-tRNA synthetase
RF	release factor
eRF	eukaryotic release factor

1 The genetic code—history

1.1 Introduction

The history of the research that led to the deciphering of the universal genetic code is interesting in view of how this most fundamental problem was solved, and how scientists were excited by this, one of the monumental achievements of this century. I must confess that I am not in an ideal position to review this history, as I played only a small part in it, and was certainly not a member of the 'RNA Tie Club' (Crick, 1968). Therefore, the next section—'The day before yesterday'—is written largely from a series of review articles by Francis Crick (1958, 1963, 1968), who has been playing the leadership-role in the coding problem, on our review article (Osawa *et al.*, 1992), and on a comprehensive review by Jukes (1977*b*).

Nowadays, everyone knows that 'DNA makes RNA makes protein'. The fundamental features of coding in this scheme are:

(1) codons (named by Sydney Brenner in 1962) are triplets of ribonucleotide;

(2) the sequences of DNA, RNA, and the amino acids in proteins are co-linear, and codons do not overlap;

(3) reading of the codons in messenger RNA starts from a fixed position, i.e. codon AUG (or sometimes GUG, etc.) and terminates at a stop codon (UAA, UAG, or UGA);

(4) codons are degenerate, i.e. there are more than two codons for most amino acids; and

(5) a codon does not interact directly with an amino acid; it is read by interaction with an amino acid-charged tRNA anticodon.

1.2 The day before yesterday

1.2.1 *Template hypothesis in the early days*

The problem as to how to determine amino acid sequence in proteins was being discussed by a few biochemists long before the DNA double helix was discovered. A naïve hypothesis was that 'the positively charged groups of basic amino acids in proteins combine with the negatively charged phosphate groups of nucleic acids, and that in this way a negative replica of the nucleic acid template is formed' (Friedrich-Freska, 1940; Jansen, 1939, cited in Haurowitz, 1950). Haurowitz (1950) could not imagine how 'a nucleic acid template could direct the different amino acids in their specific positions in the peptide chain', because 'nucleic acids consist of only seven or eight different units', and because 'combination of phosphate groups with proteins is quite non-specific'. Haurowitz and Crampton (1952) then abandoned the nucleic acid template hypothesis and suggested that amino acid sequence of proteins was determined with the sequence of the monomolecular template protein film. According to Haurowitz (1950), this idea was proposed a long time before by Muller, Haldane, and others (cited in Haurowitz, 1950). Meanwhile, evidence indicating that cytoplasmic granules (i.e. ribosomes) were the site of protein synthesis was accumulating. (The word 'ribosome' was adopted in 1958 at the 1st Symposium of the Biophysical Society (see Roberts, 1958).) Haurowitz incorporated this into his model, and stated that the surface of nucleic acid in the granules would play a role in the formation of the template protein film. Note that according to this view 'protein makes nucleic acid makes proteins', which is against the central dogma. Of course, this view is entirely incorrect, and does not explain the origin of the template, unless protein is the genetic material.

From the early 1940s to the early 1950s, biologists suspected that RNA was important in protein synthesis. Brachet (1944) and Caspersson (1950) found a close relationship between the amount of RNA and the rate of protein synthesis. (Messenger RNA was discovered much later, and the RNA with which the biologists were concerned was ribosomal RNA.)

Meanwhile, scientists began to believe that DNA was the genetic material; evidence pointing to this had been accumulating for some time. The constant amount of DNA per nucleus in a given organism, and, more importantly, DNA as the bacterial transforming principle are two examples of this evidence.

1.2.2 *The overlapping code*

In 1952 Alexander L. Dounce, at the University of Rochester, presented a clear idea of the template sequence DNA–RNA–protein. Dounce's central idea was that the amino acid sequence in a given protein was determined by the specific base sequence in the corresponding RNA, and that the information in the RNA was stored in the DNA sequence (the sequence hypothesis). Dounce, of course, did not know about the adaptor tRNA. Instead, he suggested a biochemical mechanism of direct interaction of one amino acid with every nucleotide, with the help of a set of enzymes. The amino acids were specified by the 'immediate surroundings' of an RNA base. For example, there were ten possible surroundings for A, such as AAA, GAA, CAC, etc. Ten more for each of G, C, and U gave a total of 40 triplets, which is enough to code for 20 amino acids. For example, according to Dounce, the triplet GAA would be given the same amino acid as AAG, because the direction was not considered. Dounce's code is overlapping, because the distance between adjacent two amino acids in peptide is only about 3.5 Å—the minimum distance between two adjoining bases in nucleic acid—and is therefore much shorter than that of a nucleic acid triplet. Thus, Dounce had suggested a degenerate triplet code long before the universal genetic code was established, although his scheme was not entirely correct. Dounce (1953) stated modestly:

The template hypothesis constructed by me was published in the hope of promoting thinking and experimentation of a detailed nature concerned with the possibility that nucleic acids may participate in protein synthesis.

About one year after the DNA double helix was discovered, G. Gamow (1954) at George Washington University (a well-known cosmologist and founder of the 'big-bang theory'), proposed a so-called diamond code. Gamow suggested a 'key-and-lock' relationship between various amino acids and the rhomboid-shaped holes formed by various nucleotides in the deoxyribonucleic acid molecule. He proposed 20 different holes, each of which was specific for one species of amino acid, and that 'free amino acids from the surrounding medium get caught into the "holes", and thus unite into the corresponding peptide chain'. In this model, the neighbouring holes have two common nucleotides. Therefore, this, like Dounce's model, is an overlapping code. The consequence was that 'there must exist a partial correlation between neighbouring amino acids in protein molecules'. This model was easily disproved for the following reasons: (1) there is no polarity in DNA double helix, so DNA cannot specify the direction of the peptide chain to be synthesized; (2) proteins are synthesized in cytoplasm and not in nucleus in eukaryotes; and (3) it allows limited amino acids to adjoin each other in peptides, but there is no such restriction; any amino acid can be neighbour any other. Indeed, Crick noted (1963) that 'it would not code for the known sequence of insulin'.

The triplet nature of a coding unit (codon) may be derived by simple calculations: because there are only four species of base in RNA, the minimum number of bases for a codon if 20 amino acids are to be specified is three. A doublet code would code for only 16 amino acids, but there are 64 possible triplets if there are three combinations of four bases. Overlapping codes, such as those suggested by Dounce and Gamow, are principally of a stereochemical nature, in which there is a physical relation between an amino acid and its corresponding triplet. For example, the sequence ACUAGACCUC would give eight triplets, ACU, CUA, UAG, AGA, GAC, ACC, CCU, and CUC, which could specify eight different amino acids. A characteristic of an overlapping code is that reading can start at any point without disturbing the reading frame. No commas are needed between adjoining triplets. For example, initiation from the second base, C, will give CUA, UAG, AGA, GAC, ACC, CCU, CU..., etc. Another type of overlapping code is the partially overlapping code, in which the last base of a triplet is the first base of the following triplet, e.g. ACUAGACCUC... may be read as ACU, UAG, GAC, CCU, UC..., in which the starting point must be fixed. This code also includes the restriction of neighbouring amino acids in protein sequences.

1.2.3 *The commaless code*

As there are, in fact, no restrictions on amino acid sequences in proteins, any overlapping code is unlikely to be a real one (Brenner, 1957). In addition to such restrictions, the effect of a single base substitution produced by a mutation makes it clear that the code is not overlapped. If a mutation substitutes a base U in the sequence AC$\underline{\text{U}}$AGACCUC to AC$\underline{\text{C}}$AGACCUC, the resultant arrangement of triplets becomes AC$\underline{\text{C}}$, C$\underline{\text{C}}$A, $\underline{\text{C}}$AG, AGA... in an overlapping code. Thus one base change alters three of the amino acids to be specified. Obviously, no such multiple alterations occur as the result of a single mutation. In non-overlapping (and degenerate) codes, in which RNA triplets are read consecutively without overlapping, there are no restrictions of neighbouring amino acids in protein, and a single mutation gives either no or one amino acid change, depending on whether it occurs at the silent site or at the replacement site. However, two basic difficulties with the non-overlapping code are how to choose the groups of three nucleotides and why there are 64 triplets to code for 20 amino acids. So, for example, the sequence ...UCACGGAUAUGC... may be read as ...UCA, CGG, AUA, UGC..., ...U, CAC, GGA, UAU, GC... or UC, ACG, GAU, AUG, C..., depending on the starting point. If the starting point is fixed, there is no problem but at that time (1957), the existence of initiation codon was not known.

Crick *et al.* (1957) offered a solution to this problem by proposing that 'certain triplets make sense and some make nonsense', so that the 64 codons may be reduced to a total of 20, one for each amino acid. Such a code may be written as:

$$\mathrm{AB}\begin{array}{c}\mathrm{A}\\ \mathrm{B}\end{array}\qquad \begin{array}{c}\mathrm{A}\\ \mathrm{B}\end{array}\mathrm{C}\begin{array}{c}\mathrm{A}\\ \mathrm{B}\\ \mathrm{C}\end{array}\qquad \begin{array}{c}\mathrm{A}\\ \mathrm{B}\\ \mathrm{C}\end{array}\mathrm{D}\begin{array}{c}\mathrm{A}\\ \mathrm{B}\\ \mathrm{C}\\ \mathrm{D}\end{array}$$

where $\mathrm{AB}\begin{array}{c}\mathrm{A}\\ \mathrm{B}\end{array}$ donotes ABA and ABB, etc.

This code implies that 'no sequence of these allowed triplets will ever give one of the allowed triplets in a false position, so as to automatically remove commas (a commaless code) without fixing the starting point' (Crick 1958). This was a most elegant theoretical solution to the coding problem, and yet had problems of its own. As the commaless code is a non-degenerate code, there is only a limited flexibility against mutations. Any single mutation in a gene produces either a change in amino acid assignment, or, more seriously, a nonsense codon that inactivates the whole gene in most cases, so that the mutational load is quite high. In fact, 70 per cent of possible single nucleotide substitutions will give rise to nonsense codons. From the evolutionary point of view, the flexibility is recognized among homologous genes in different organisms with different genomic G + C contents. The silent sites of such genes can become higher in G + C without changing the amino acid sequence along with the increasing genomic G + C content. No such flexibility exists in the commaless code, and therefore evolution is virtually impossible.

Crick (1966*b*) notes that:

> This turned out to be one of those nice ideas which is, nevertheless, completely wrong. We ourselves began to lose faith in it when we eventually noticed the wide range of DNA

compositions which occur in various micro-organisms. . . . Unfortunately, people found the idea so pretty that it was widely referred to, and even found its way into a popular book on the subject. Personally, I was always very undecided about it.

1.2.4 *The adaptor hypothesis*

At about the same time, Crick (1955, cited in Hoagland, 1960; 1957) proposed an ingenious idea about how to locate the amino acids in RNA sequence, which he called the adaptor hypothesis. Crick states:

I cannot conceive of any structure (RNA or DNA) acting as a direct template for amino acids, or at least as a specific template. In other words, if one considers the physico-chemical nature of the amino acid side chains we do not find complementary features on the nucleic acid . . . What the DNA structure does show (and probably RNA will do the same) is a specific pattern of ***hydrogen bonds***, and very little else . . . Each amino acid would combine chemically, at a special enzyme, with a small molecule which, having a specific hydrogen-bonding surface, would combine specifically with the nucleic acid template. This combination would also supply the energy necessary for polymerization. In its simplest form there would be 20 different kinds of adaptor molecule, one for each amino acid, and 20 different enzymes to join the amino acid to their adaptors. (Crick, 1955; cited in Hoagland, 1960)

The adaptor was assumed to be a tri- or somewhat longer polynucleotide attached to the amino acid. The base sequence of the adaptor would be complementary to that of the triplet (codon) and could pair with it by hydrogen bonding, so that the amino acid can find the correct place on the template. Crick's adaptor proved to be the much larger molecule now known as tRNA. Although the adaptor hypothesis was proposed to apply to the commaless code, the hypothesis itself is, in its essence, valid for the decoding of the genetic code now known. Among numerous 'pure' theories of the coding problems (which were mostly incorrect, as seen above), the adaptor hypothesis is the most remarkable example to be verified experimentally.

1.2.5 *tRNA as an adaptor*

In the mid-1950s, Zamecnik, Hoagland, and their co-workers at Harvard Medical School, initiated comprehensive studies on the biochemical mechanism of protein synthesis *in vitro*, mainly using rat liver. They found that the amino acid was first activated to form aminoacyl adenylate, which was then transferred to a specific RNA molecule, now known as tRNA. These two processes were catalysed by the same enzyme. Hoagland (1960) clearly stated that 'twenty amino acids use twenty enzymes to convey them to twenty specific RNA molecules' (Hoagland, 1960; Zamecnik, 1969). (At that time, of course, isoacceptor tRNAs were not known.) The amino acids so bound to tRNA were subsequently transferred to the template RNA on the ribosomes, where the condensation of amino acids took place. The complementary interaction of tRNA with template RNA was suggested. This concept was based on experimental evidence and was arrived at independently from the adaptor hypothesis, although the two are complementary. Zamecnik *et al.* thought that the template was ribosomal RNA, which, however, does not have enough information for the large variety of protein species in an organism, unless ribosomal RNAs are heterogeneous enough. It was some time before messenger RNA was discovered.

1.2.6 *Triplet nature and fixed reading frame of the code*

In 1961, Crick *et al.* found strong evidence to suggest that the code is read in triplets starting from a fixed point, so that the phase of reading is defined. This finding stemmed from extensive genetic experiments using the rII region of the T4 bacteriophage. A frame-shift mutant, FCo, in the B cistron of the rII region was obtained by proflavin treatment. The mutant could form an atypical plaque in *E. coli* B but not at all in the K strain. Further treatment of the mutant by proflavin gave revertants that could form a plaque in the K strain. The genetic analyses suggested that the mutation in FCo was caused by a single base addition (+) and that in the second mutants, one base deletion (−) occurred near the (+) region, thus reverting to the wild-type by cancelling the first mutation. This second mutation was called a suppressor mutation. Construction of double mutants by recombination of FCo with suppressors, identified eight suppressors. They further obtained the suppressors (+) of suppressors (−) and, by recombination of these mutants, various double and triple mutants. These mutants revealed the following characteristics:

1. rIIB cistron was not active either in (+) or (−) alone, but active in (+, −) or (−, +).
2. It was inactive in (+, +) and (−, −).
3. It was active in (−, −, −) or (+, +, +) but inactive in (+, +, −) or (+, −, −).

The results were interpreted as follows: When the first mutation was (+) or (−), the correct reading frame was shifted to downstream or to upstream by one base. If the second mutation—(−) or (+)—existed close to the first mutation site, the reading frame was corrected thereafter. This would result in altered amino acids in the limited region, which, however, did not seriously affect the rIIB activity. Thus it became clear that the reading of the code started from a fixed point, and must have included commas. In the commaless code, (+, −) or (−, +) mutations would produce a number of 'nonsense' codons that stopped protein synthesis. In the triple mutants (+, +, +) or (−, −, −), the reading frame was supposed to recover by the addition or deletion of three bases (or some higher multiple of three; this is not the case, as we now know). These triple mutations would cause insertion or deletion of one amino acid, without much effect on the phenotype.

These results implied that the code was read by the triplet-phase, and Crick *et al.*'s conclusions were later verified by amino acid changes in mutants of human haemoglobin, tobacco mosaic virus (TMV), and phage T4 lysozyme.

1.2.7 *The discovery of messenger RNA*

Volkin and Astrachan (1956) demonstrated the synthesis of a new class of RNA after phage T2 infection of *E. coli*. The new RNA revealed a high turnover rate and had a base composition similar to that of phage DNA. Brenner *et al.* (1961) showed that this RNA was not rRNA, but was transcribed from phage DNA, then combined with pre-existing *E. coli* ribosomes, and acted as the template for the synthesis of phage proteins. This is the first clear demonstration of messenger RNA, which was later found unequivocally in all organisms. This discovery solved the problem of the information content of template RNA.

1.3 Yesterday

1.3.1 *The breakthrough*

Despite numerous theoretical and experimental approaches to the coding problem and to the mechanism of protein synthesis, it was not until 1960 that any codons were identified. The breakthrough was brought about by the well-known work by Nirenberg and Matthaei (1961), at the National Institute of Health, USA, who found that poly-U and C directed the synthesis of polyphenylalanine and polyproline, respectively, in an *E. coli* cell-free protein-synthesizing system. Their obvious conclusion was that UUU and CCC were, respectively, codons for Phe and Pro. AAA was later found to be a codon for Lys. Incidentally, this finding disproved the commaless code, as both UUU and CCC must have been nonsense codons in this code. Using Nirenberg and Matthaei's findings, Ochoa's group (Speyer *et al.*, 1963), at New York University, extended analyses using synthetic ribopolymers of known composition but with a random sequence. For example, poly-AC containing A and C in the ratio 5 : 1 stimulated the incorporation of six amino acids. The amount of incorporated amino acid relative to Lys made it clear that codon composition for Asp and Gln was 2A1C; His, 1A2C;Pro, 1A2C and 3C; Thr, 2A1C and 1A2C. In this way, the codon compositions for all 20 amino acids were elucidated by the two American groups. However, these experiments did not allow the determination of the sequence of codons, except for UUU, AAA, CCC, and GGG.

1.3.2 *Deciphering the amino acid code*

The second breakthrough was again accomplished by Nirenberg's group. Nirenberg and Leder (1964) succeeded in binding a trinucleotide of known sequence with *E. coli* ribosomes: aminoacyl-tRNA with complementary base sequences bonds specifically with the corresponding trinucleotide *in vitro* (trinucleotide-stimulated ribosomal binding reaction).In this way, the codon sequences for Leu, Cys, and Val with 2U1G composition, were determined to be UUG, UGU, and GUU, respectively.

At about the same time. Khorana's group (Nishimura *et al.*, 1965), at the University of Wisconsin, succeeded in preparing various polynucleotides composed of two different bases arranged alternately. In an *in vitro* translation system, poly-UC (...UCUCUCUCUCUC...) gives a polymer of ...Ser–Leu–Ser–Leu..., and poly-AG (...AGAGAGAGAG...) gives ...Arg–Glu–Arg–Glu–Arg.... Thus, it became clear that UCU is a codon for Ser, CUC for Leu, AGA for Arg, and GAG for Glu. The experimental work carried out by Nirenberg and Khorana's groups, assigned codons for 61 out of a possible 64 triplets to 20 amino acids, as now can be seen in the universal genetic code table.

1.3.3 *Termination codons*

In Nirenberg's binding assay, the trinucleotides UAA, UAG, and UGA did not bind any aminoacyl-tRNA. These were therefore suspected to be signals for termination. Takanami and Yan (1965) were the first to show that only synthetic polynucleotides containing U and A caused a significant release of peptides in the supernatant fraction when tested in a cell-free amino acid-incorporation system from *E. coli*. Khorana *et al.* (1966) identified UAA and UAG as termination codons by examining amino acid incorporations stimulated by messengers containing repeating nucleotide sequences of UAGA or GUAA. In every fourth position, these two polynucleotides contain UAG and UAA, respectively. No polypeptide synthesis was observed with either messengers, in contrast to the significant synthesis with the other repeating polynucleotides that were examined.

A termination codon was also identified in a 'nonsense' mutant of the T4 phage, in which a certain amino acid codon in the phage head protein mutated to another codon not corresponding to any amino acid (Brenner *et al.*, 1965). In this mutant, protein synthesis ceased at the mutant codon, so that an immature peptide chain was released from the ribosomes and the phage could not grow. Brenner *et al.* isolated a series of mutant phage that could grow in a suppressor strain (Su^+) but not in the Su^- (wild-type) strain. The direct analysis of the amino acid sequence of the head protein showed that some of the protein from phage grown in the Su^+ *E. coli* had Ser at the Gln site in the wild-type protein. By using other Su^+ strains of a similar nature (collectively called amber suppressors), substitutions of Gln by other amino acids were found. These and other experimental results, together with the subsequent characterization of suppressor tRNAs from the Su^+ strains, indicated that a wild-type Gln codon was substituted by a termination codon in the 'nonsense' mutant. This termination codon could be read by suppressor tRNAs for amino acids in the Su^+ *E. coli* strains. For example, the suppressor $tRNA^{Ser}$ with the anticodon sequence CUA (wild-type CUG) is responsible for reading the 'nonsense' codon as Ser in the Su^+ *E. coli*, indicating that the wild-type Gln codon was CAG and was changed to the termination codon UAG by a single C-to-U mutation in the mutant phage. Similar experiments with other types of Su^+ strains (ocher and opal suppressors) showed that UAA and UGA were also termination codons (Stretton *et al.*, 1966). The suppressor tRNA is derived from the wild-type tRNA. It usually has a single base substitution in its anticodon, so that it can pair with the corresponding termination codons UAA, UAG, or UGA. The assignment of these three codons to the termination signal was confirmed by other experiments which are not mentioned here.

1.3.4 *Initiation codons*

As protein synthesis starts from a fixed point in messenger RNA, there must be a codon(s) for initiation. In 1964, Marcker and Sanger discovered that *N*-formylmethionyl-tRNA was formed in *E. coli*, and suggested that this tRNA was involved in protein chain initiation. Webster *et al.* (1966) found that the *N*-terminal amino acid residue for all the proteins synthesized *in vitro* with phage RNA as the template was formylmethionine (fMet). At about the same time, Clark and Marcker (1966) identified two species of $tRNA^{Met}_{CAU}$ in *E. coli*. One of these was charged by the fMet that was incorporated only in the *N*-terminus of proteins. These experiments suggested that the initiation codon was AUG, which is also a codon for Met. Subsequently, elegant experiments by Steitz (1968) showed that the initiation codon for R17 RNA phage proteins was indeed AUG. It is now known that other codons, notably GUG, are used for initiation. However, GUG, which is a codon for Val, is recognized by $tRNA^{fMet}_{CAU}$ when it is at the initiation site. Throughout this book, the three-letter amino acid symbol and nucleotide triplet following 'tRNA' denotes the tRNA identity (i.e. the amino acid to be accepted by this tRNA, and the anticodon of the tRNA, respectively.)

1.3.5 *The genetic code* in vivo

As described above, the amino acid assignment of 61 codons was determined *in vitro* using synthetic polynucleotides and trinucleotides as messengers. However, synthetic polynucleotides do not have initiation and termination signals, and therefore the peptide synthesis starts and stops at arbitrary points. There was therefore no guarantee that the *in vitro* code *per se* worked *in vivo*.

Streisinger's group (Terzaghi *et al.*, 1966; Okada *et al.*, 1966) was the first to demonstrate that the *in vivo* code was the same as the *in vitro* code. They compared the amino acid sequence of lysozyme from wild-type T4 phage with that of a double mutant phage (prepared by crossing two frame-shift mutants). Analysis of these considered all the possibilities that could be deduced from the *in vitro* amino acid code. A similar conclusion was also drawn from analyses of a series of TMV mutants.

The most elaborate and conclusive work on the identity of the *in vivo/in vitro* code was performed by Fiers *et al.* (1976), who sequenced MS2 phage proteins and aligned these with the sequence of the viral RNA. All the codes deduced from the *in vitro* experiments with synthetic messengers proved to be correct. Also, the large number of variant haemoglobins did a lot to establish the identity of the *in vitro* and *in vivo* genetic codes. Thus, the genetic code in Table 1.1, which is degenerate and is not comma-free, was established without any doubt. The code was thought to be 'universal', because it was common among *E. coli*, certain other organisms and viruses. This led Crick (1968) to propose the frozen-accident

Table 1.1 The universal genetic code

Codon	Amino acid		Codon	Amino acid		Codon	Amino acid		Codon	Amino acid	
UUU	Phenylalanine	(Phe)	UCU	Serine	(Ser)	UAU	Tyrosine	(Tyr)	UGU	Cysteine	(Cys)
UUC	Phenylalanine		UCC	Serine		UAC	Tyrosine		UGC	Cysteine	
UUA	Leucine	(Leu)	UCA	Serine		UAA	Stop		UGA	Stop	
UUG	Leucine		UCG	Serine		UAG	Stop		UGG	Tryptophan	(Trp)
CUU	Leucine	(Leu)	CCU	Proline	(Pro)	CAU	Histidine	(His)	CGU	Arginine	(Arg)
CUC	Leucine		CCC	Proline		CAC	Histidine		CGC	Arginine	
CUA	Leucine		CCA	Proline		CAA	Glutamine	(Gln)	CGA	Arginine	
CUG	Leucine		CCG	Proline		CAG	Glutamine		CGG	Arginine	
AUU	Isoleucine	(Ile)	ACU	Threonine	(Thr)	AAU	Asparagine	(Asn)	AGU	Serine	(Ser)
AUC	Isoleucine		ACC	Threonine		AAC	Asparagine		AGC	Serine	
AUA	Isoleucine		ACA	Threonine		AAA	Lysine	(Lys)	AGA	Arginine	(Arg)
AUG	Methionine	(Met)	ACG	Threonine		AAC	Lysine		AGG	Arginine	
GUU	Valine	(Val)	GCU	Alanine	(Ala)	GAU	Aspartic acid	(Asp)	GGU	Glycine	(Gly)
GUC	Valine		GCC	Alanine		GAC	Aspartic acid		GGC	Glycine	
GUA	Valine		GCA	Alanine		GAA	Glutamic acid	(Glu)	GGA	Glycine	
GUG	Valine		GCG	Alanine		GAG	Glutamic acid		GGG	Glycine	

theory, which stated that evolution of the code had stopped accidentally in the progenote from which all the present-day organisms descended (see Section 6.1).

1.3.6 *The wobble hypothesis*

When we look at the genetic code table, most of the first and all of the second bases of codons for one amino acid are clearly distinguished from others, while at the third position, U and C, and A and G are equivalent (in the family boxes, all four bases are equivalent). Crick (1966*a*) suggests that the pairings of the first and second positions with tRNA anticodon are likely to be the standard ones, i.e. G pairs with C and A with U, and 'the pairing in the third position might be *close* to the standard ones'. By examining the possible base pairs close to the standard, Crick proposed the wobble hypothesis, stating that: 'in the base-pairing of the third base of the codon (with anticodon first base) there is a certain amount of play, or wobble'. For example, U at the first anticodon position pairs with A as well as G at the third codon position, etc. (Table 1.2). The non-overlapping degenerate triplet nature of the genetic code is a result of these rules. The currently revised wobble rules will be described in Section 2.2.

Table 1.2 Codon–anticodon pairing rules

Anticodon	Codon
U	A G
C	G
A	U
G	U C
I	U C A

From Crick (1966*a*).

1.3.7 *tRNA*

As noted above, Zamecnik *et al.* (see Zamecnik, 1969, for review) were the first to discover an RNA corresponding to the adaptor molecule. This RNA was called sRNA (soluble RNA), as it was left in the supernatant fraction after ribosomes were sedimented by ultracentrifugation. In the early 1960s, the name transfer RNA (tRNA) replaced sRNA, because the RNA transfers amino acids to the template RNA. That tRNA is indeed the adaptor was demonstrated most elegantly by Chapeville *et al.* (1962). Cys attached to $tRNA^{Cys}$ was converted to Ala by reduction with Raney Nickel. The resultant Ala attached to $tRNA^{Cys}$ was incorporated into peptide with poly-UG as a messenger, indicating that UGU (a Cys codon) was read as Ala instead of Cys by recognizing the $tRNA^{Cys}$ rather than $tRNA^{Ala}$.

There must be at least 20 species of tRNA for 20 amino acids and Crick and Zamecnik suggested the existence of aminoacylation enzyme (aminoacyl tRNA synthetase; ARS) corresponding to each amino acid. These were verified by the fractionation of tRNA and ARS, by various procedures and by a number of workers. It was also found that there was generally more than one species of tRNA for a single amino acid (isoacceptors).

The primary structure of tRNA was first determined by Holley and co-workers (1965) for $tRNA^{Ala}$ from yeast. Three possible secondary structures were proposed, one of which is now known as the clover-leaf structure common to all the tRNA species (Fig. 1.1a) except some mitochondrial tRNAs. The model clearly had the anticodon sequence IGC in the loop structure. This provided a basis for I pairing with U, C, and A in the codon first position in the wobble hypothesis. There are now known to be hundreds of tRNA structures in many different organisms and organelles. All the predicted 54 anticodons of tRNA (Jukes, 1977*a*) have now been identified, except ICC, which probably does not exist. Note that there are 56 possible anticodons, including UUA, CUA (in ciliates), and UCA (in *Mycoplasma* and mitochondria) (see Section 3).

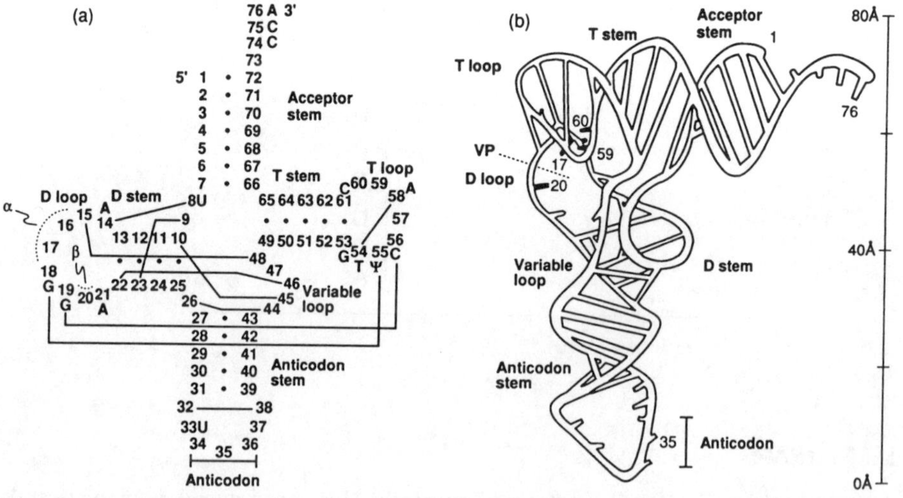

Fig. 1.1. Structures of tRNA. (a) Cloverleaf structure of tRNA with the Cold Spring Harbor numbering system of nucleotides in tRNA (Schimmel, Söll, and Abelson, 1979). Constant nucleotides are noted. ● Watson–Crick base pairings; —— other base pairings. (b) L-shaped structure of tRNA (from McClain, 1993).

From the late 1960s, X-ray diffraction studies of the tertiary structure of tRNA began. In 1974, independent groups headed by Rich (Kim *et al.*, 1974*a*,*b*) and Klug (Robertus *et al.*, 1974) succeeded in analysing (at 3 Å resolution) yeast $tRNA^{Phe}$, and presented its L-shaped tertiary structure. The model was subsequently proved to be common for all tRNAs (Fig. 1.1b), except some mitochondrial tRNAs. The discovery of the L-shaped structure stimulated studies on the interaction between tRNA and ARS, and on conformational changes in codon–anticodon interactions, and so on.

2 The structure of the 'universal' genetic code

2.1 Codons

The genetic code is written as a set of all 64 possible arrangements of the four RNA nucleotides—U, C, A, and G—in the form of a triplet (codon) (see Table 1.1). Each of 61 codons is assigned to a specific amino acid; the other three are stop codons that terminate the protein synthesis and do not correspond to any amino acid. As the number of amino acids used for protein synthesis is 20, the same amino acid can be coded for by more than one codon, i.e. the code is degenerate. For example, AAA and AAG are codons for Lys; GUG, GUC, GUA, and GUG are codons for Val. Further, six codons—CGU, CGC, CGA, CGG, AGA, and AGG—are synonymous, and are translated to Arg. A box in which four codons are synonymous is called 'family box' or a '4-codon box'; there are eight family boxes. A set in which a single amino acid has two codons is called '2-codon set' (see Table 1.1). Of the 13 2-codon sets, seven end with pyrimidine (Y: U or C) and six with purine (R: A or G). Codons for Arg, Leu, and Ser exist both in a family box and in a 2-codon set. There are exceptions: three codons—AUU, AUC, and AUA—are assigned to Ile (2-codon set plus 1, or family box minus 1). Met or Trp has only a single codon—AUG or UGG—because codon AUA is for Ile and UGA is a stop codon. There are therefore two to six synonymous codons for a single amino acid, except for Met and Trp.

In each set of synonymous codons, the first and the second nucleotides are common, and two to four of the third nucleotides are free to change without changing the amino acid assignment. Such a 'free site' is extended to the first nucleotide in the case of Arg and Leu. In family boxes, the third nucleotides are all free to change, while in the 2-codon sets, possible changes exist only between A and G, or U and C of the third nucleotide. These sites are often collectively referred to as silent sites. However, such sites in 2-codon sets are actually semi-silent, because in codons for Lys, for example, the freedom exists only between AAA and AAG, and a change from AAA to AAU or to AAC results in a change of amino acid assignment from Lys to Asn. Only transitions are silent in 2-codon sets.

A site where the change is accompanied by a change of amino acid assignment is called a 'replacement site'; the second nucleotide in all codons and the first nucleotide in all except Leu and Arg are replacement sites. Mutations occur at all the codon sites randomly. A small number of mutations occurring at the silent sites is fixed in a population by random genetic drift; mutations at the replacement sites are removed by negative selection if the resulting amino acid change is deleterious. When the amino acid change is only slightly or not at all deleterious, the mutation may be fixed, as in the case of the silent mutation. Such a mutation, called 'conservative mutation', often takes place between codons coding similar amino acids, such as Lys and Arg (AAR↔AGR), Asp and Asn (GAY↔AAY), Gln and Glu (CAR↔GAR), Ile and Val (AUY/A↔GUN; N = A, G, U, or C), or Ile and Leu (CUN↔AUY). It must be remembered that mutations and substitutions are different. What we actually observe for any codon change in natural populations is a substitution resulting from fixation of a mutation, as noted above. The mutation rate is therefore much higher than the substitution rate, although the higher the former, the higher the latter (Kimura, 1983).

2.2 Codon–anticodon pairing rules

The genetic code consists of another essential component—the anticodon of the tRNA molecule. Each tRNA has molecular weight of about 25 000, consists of about 80 nucleotides, and has an anticodon sequence of three nucleotides at positions 34, 35, and 36 in the anticodon loop. The tRNAs are 'adaptors' that transfer amino acids to the specific codons in mRNAs during protein synthesis on the ribosomes. Each tRNA molecule is characterized by its anticodon and by the attachment of a specific amino acid to its 3′-terminal adenine (the tRNA identity). Anticodons and codons pair by hydrogen bonding. Pairing between the second and the third positions of an anticodon, and the second and the first positions of a codon, follows the Watson–Crick-type pairing rules, i.e. A pairs with U, and G pairs with C. There is some ambiguity, or wobble between the first anticodon nucleotide and the third codon nucleotide upon pairing. For example, G (anticodon) pairs with U as well as C (codon), so that all pairs of codons ending with a pyrimidine, e.g. UUU and UUC, are translated to Phe by a single anticodon—GAA. An anticodon AAA for codon UUU does not exist.

In his original wobble hypothesis, Crick (1966*a*) proposed that G (anticodon) pairs with U and C (codon), U with A and G, and C only with G. A pairs with U, but A is always modified to I (inosine, the base hypoxanthine) and pairs with U, C, and A (see Section 1.3.6).

Between 1965 and 1970, these rules were examined experimentally by studying the coding specificity of isolated tRNA species in the trinucleotide-stimulated ribosomal binding reaction, or in amino acid incorporation assay performed with ribopolynucleotide messengers of known base composition (Matthaei *et al.*, 1966; Nirenberg *et al.*, 1966; Söll *et al.*, 1966*a,b*). The results were generally in agreement with the wobble hypothesis.

The original wobble rules have expanded, mainly as a result of the finding that the first nucleotide of the anticodon is modified in various ways, and the codon recognition is greatly influenced by such modifications. The latest version of the wobble rules is given in Table 2.1.

Table 2.1 Current status of codon–anticodon pairing rules

First anticodon nucleoside	Third codon nucleoside	Usage	Systems
U	U, C, A, G	Family boxes	Mitochondria, *Mycoplasma* spp., chloroplasts
xo^5U (^+U)[a]	U, A, G	Family boxes	Eubacteria
$cmnm^5U$, mcm^5U, Um ($^*U$)[a]	A, G	2-codon sets	Mitochondria, bacteria, eukaryotes
xm^5s^2U ($^*U$)[a]	A, (G)	2-codon sets	Eubacteria, eukaryotes
G	U, C	2-codon sets	All
G	U, C	Family boxes	Bacteria
Q	U, C	2-codon sets	Eubacteria, eukaryotes
I	U, C, A	Arg CGN	Eubacteria
I	U, C, A	All family boxes except Gly GGN	Eukaryotes
A (rare)	U, C, G>A	Thr ACU, Arg CGN	*Mycoplasma* spp., yeast mitochondria
C	G	All	All
f^5C	Probably A, G	Met AUR	Animal mitochondria
L (2-lysyl C)	A	Ile AUA	Eubacteria, plant mitochondria

From Osawa *et al.* (1992), with modifications. [a] Abbreviations used in text.

2.2.1 *G*

In many bacteria and eukaryotes, G at the first anticodon position of tRNA that recognizes NAY in a 2-codon set is modified to queuosine (Q) or its derivative (Nishimura, 1979, 1983). $tRNA^{Tyr}$ with G, instead of Q, at the first anticodon position often mistakenly pairs *in vitro* with the stop codon UGA (Nishimura, 1983). It is possible that Q ensures recognition of the codon NAY, thus preventing mispairing with the codon NAR for another amino acid (or stop codon). However, there are exceptions. The first anticodon nucleoside G for NAY codons in *Micrococcus luteus* (Kano *et al.*, 1991), *Mycoplasma capricolum* (Andachi *et al.*, 1989), and some mitochondria (Sprinzl *et al.*, 1989) is not modified. The misreading of codon NAR by anticodon GUN might be prevented in some other way, such as by the influence of sequences outside the anticodon, or it might occur at such a low frequency that appreciable functional disturbance does not occur.

The first anticodon nucleoside G is unmodified in tRNAs for codons in the bacterial family boxes. The misreading of codon NAR by anticodon GUN, if it occurs, is not harmful, because NAR and NAY code for the same amino acid in family boxes. As described below, in eukaryotes, I (inosine) replaces G in all family boxes except Gly.

2.2.2 *U*

Crick (1966*a*) noted that U : U and U : C base-pairs were possible, but dropped them from his wobble rules because they were too close and would cause misreading in 2-codon sets.

In 1981, the human mitochondrial genome was sequenced (Anderson *et al.*, 1981). As a result, it became apparent that there is only one tRNA species per family box in human mitochondria (Barrell *et al.*, 1980). Heckman *et al.* (1980) showed that the family box tRNAs from *Neurospora crassa* mitochondria have an unmodified U at the first anticodon position, and suggested that all four codons in a family box are read by the UNN anticodon. Evidence to support this was obtained from yeast mitochondria (Bonitz *et al.*, 1980). This form of codon–anticodon structure was later found in family boxes of most of the non-plant mitochondria, *Mycoplasma* spp. (Andachi *et al.*, 1987, 1989; Samuelsson *et al.*, 1987), and in two family boxes of the chloroplast code (Shinozaki *et al.*, 1986; Umesono and Ozeki, 1987).

Samuelsson *et al.* (1987) showed that *Mycoplasma mycoides* $tRNA^{Gly}_{UCC}$, which is the sole $tRNA^{Gly}$ in this bacterium, can read at least three Gly codons—GGU, GGC, and GGA—when tested in an *in vitro* protein synthesizing system using MS2-RNA as the message (codon GGG does not exist in the tested message). Kojima *et al.* (unpublished data) showed that with synthetic mRNA containing the test codons, *Mycoplasma capricolum* $tRNA^{Thr}_{UGU}$ and $tRNA^{Ala}_{UGC}$ (both unmodified U at the wobble position) can read all four codons in the respective family box in the cell-free translation system (Fig. 2.1). Thus, an unmodified U at the first anticodon position apparently pairs with all four bases—U, C, A, and

G—in the third position of codons (4-way wobble), although Lagerkvist (1978, 1981) gave a different explanation (the two-out-of-three mechanism; see page 25). Lustig *et al.* (1993) found that a mutant *E. coli* $tRNA^{Gly}_{UCC}$ (U, unmodified) reads GGA and GGG codons but not GGU and GGC. When U32 of this tRNA is replaced by C32, the tRNA reads all four Gly codons, suggesting that nucleotide 32, in addition to unmodified U at anticodon first position, influences the reading properties of the anticodon UCC. Grosjean *et al.* (1978) state that, in addition to the long wobble pairs U : A and U : G the short wobble pairs U : C and U : U are streochemically permitted, providing an adjustment is made in the polynucleotide back chain. The misreading of C or U (codon) by U (anticodon) in 2-codon sets is not a problem because U is always modified to *U so as to pair only with A and G, as described below.

```
pppGGGCGAAUUGGAGCUC AGGAGG GUUAC
                    <S.D.>
AUG AUCAUCAUCAUCAUCAUCAUCAUUAUUAUUXXX XXX UACUACUACUACUAC UAA
Met <----------10 x Ile---------->          <---5 x Tyr---> Stop
                                       CUGCAGGAAUUCGAUAUCAAGCU-OH
```

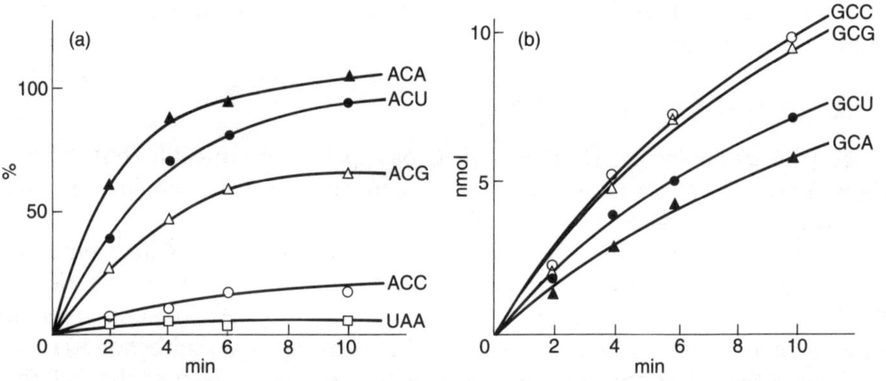

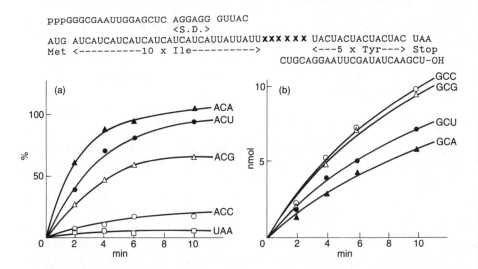

Fig. 2.1. Anticodon UNN reads all four codons in a family box. Incorporation of [^{3}H]radioactivity from (a) [^{3}H]threonyl-$tRNA^{Thr}_{UGU}$ and (b) [^{3}H]Ala into peptides was measured with the S30 fraction of *Mycoplasma capricolum.* [^{3}H]incorporation was expressed as percent of 'in-put' radioactivity of [^{3}H]threonyl-$tRNA^{Thr}_{UGU}$ for (a). [^{3}H]Ala was added directly into the reaction mixture in experiment for (b), because only one species of $tRNA^{Ala}$ with anticodon sequence UGC exists in *M. capricolum*. The mRNA used is shown above the figures, in which a pair of test codon indicated in each figure is expressed as XXX. S.D., Shine–Dalgarno sequence (from Kojima *et al.*, unpublished).

U at the anticodon first position is modified in most of the coding systems throughout prokaryotes and eukaryotes, and also in 2-codon sets of mitochondria (see page 19). Various modifications of U restrict its pairing. One modification of U, +U (xo^{5}U), is a derivative of uridine 5-oxyacetic acid that has been found in bacterial family boxes (Fig. 2.2). This modification was reported to recognize the U of the third nucleoside of codon at some 20 per cent of its recognition level

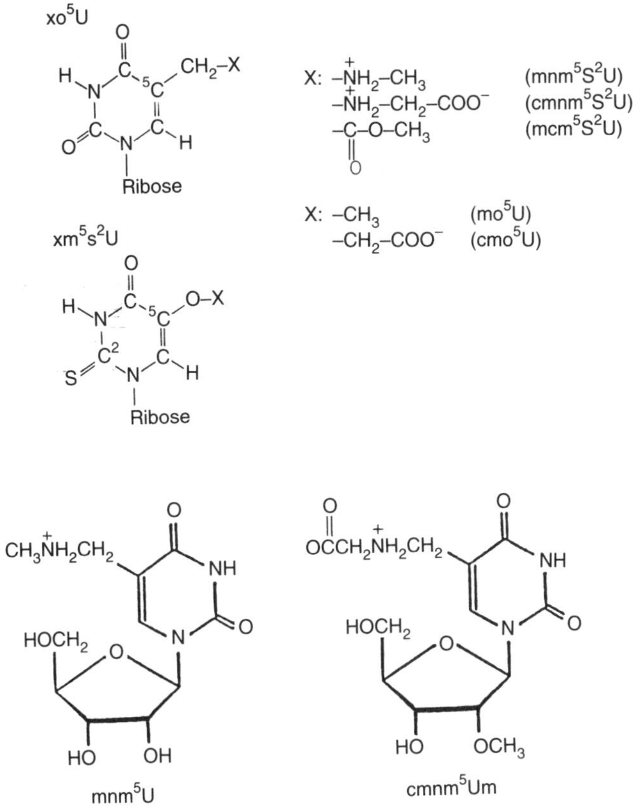

Fig. 2.2. Structures of various modified uridine derivatives found in the first anticodon position. xo^{5}U pairs with A, G, and U of the codon third nucleoside. Other uridine derivatives pair mainly with A and, to a lesser extent, with G (from Yokoyama *et al.*, 1985, for xo^{5}U and xm^{5}s^{2}U; and S. Yokoyama for mnm^{5}U and cmm^{5}Um (courtesy of S. Yokoyama)).

of A and G (Harada and Nishimura, 1972). The xo^{5}U takes the c$^{3'}$-endo form as well as the c$^{2'}$-endo form, and is thus flexible enough to recognize U, as well as A and G, at the third position of codons (Yokoyama *et al.*, 1985). Presumably, unmodified U is also flexible in this respect, and so can pair with all four bases.

Unmodified U, or +U at the first anticodon position, occurs only in family boxes and never in 2-codon sets. NNY and NNR in 12 2-codon sets code for different amino acids. Therefore U at the wobble position of the anticodon for NNR codons in 2-codon sets must be modified to a form other than +U, so as to prevent it from mispairing with NNY codons. Heckman *et al.* (1980) showed that, in mitochondrial tRNAs that read NNR 2-codon sets, the U in this position is always modified, while that in family boxes is not modified. Ohashi *et al.* (1970)

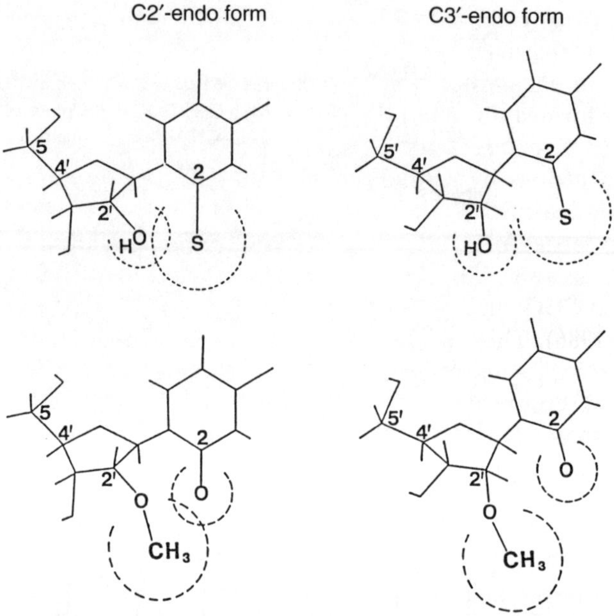

Fig. 2.3. Intramolecular interaction in 2-thiouridine derivative and in 2′-O-methyl pyrimidine derivative. Steric effect prevents to take the C$^{2'}$-endo form and stabilizes the C$^{3'}$-endo form (courtesy of S. Yokoyama).

and Yoshida *et al.* (1970) first found that in tRNA^{Glu} of *E. coli* and tRNA^{Glu} of yeast, the first anticodon nucleoside is a thiouridine derivative probably 5-acetyl-amino-methyl-2-thiouridine in *E. coli*, and 5-acetyl-methyl-2-thiouridine in yeast. In the first anticodon position of tRNA species with a methylthiouridine derivative (xm^{5}s^{2}U*U) (see Fig. 2.2), the steric effect between the 2-thiocarbonyl group and 2′-hydroxy group stabilizes the c$^{3'}$-endo form by preventing it from taking the c$^{2'}$-endo form, and enhances the rigidity of the anticodon moiety to allow a stable pairing with A at the third position of the codon (Yokoyama *et al.*, 1985) (see Fig. 2.3). The modified forms cmnm^{5}U, mcm^{5}U, and Um*U, although lacking the 2-thiocarbonyl group, take the c$^{3'}$-endo form and pair somewhat more efficiently with A than with G, so that the misrecognition of codons terminating with uridine is prevented (Martin *et al.*, 1990). Because of weak, or almost no, pairing of *U with G, NNR 2-codon sets are often translated by two anticodons—the G-terminated codon mainly by anticodon CNN, and the A-terminated codon by anticodon *UNN.

In prokaryotes, anticodon xm^{5}s^{2}UNN must translate codon NNG to some extent. For example, AAR Lys or GAR Glu 2-codon sets in *E. coli* contain only a single tRNA species with anticodon *UUU or *UUC, and yet codon AAA or GAA is used at 10 to 30 per cent of the level in AAR or GAR codons. In fact, early works by Söll *et al.* (1966*a*,*b*) and Caskey *et al.* (1968), using trinucleotide-dependent binding of aminoacyl-tRNA to ribosomes, demonstrated that AAR, GAR, or AGR (Arg) codons could be recognized by a single species of tRNA (later identified as that having anticodon *UUU, *UUC, or *UCU).

The situation in eukaryotes seems to be somewhat different. Söll *et al.* and Caskey *et al.* performed the binding experiments mentioned above using purified yeast tRNAs and *E. coli* ribosomes, and demonstrated the recognition of both NNA and NNG codons by a single anticodon, *UNN. However, with similar experiments using yeast tRNAs and yeast ribosomes, the results were obscure. The eukaryotic code carries a full complement of 15 CNN anticodons. For example, *Saccharomyces cerevisiae* contains two species of genes for $tRNA^{Gln}$ with anticodons UUG and CUG, and the deletion of the gene for $tRNA^{Gln}_{CUG}$ is lethal (Weiss and Friedberg, 1986). This would indicate that the Gln CAG codon is normally translated by $tRNA^{Gln}_{CUG}$ and not (at least not efficiently) by $tRNA^{Gln}_{UUG}$. Guthrie and Abelson (1982) proposed that, at least in yeast, the codons NNA and NNG are recognized separately by anticodons *UNN and CNN, respectively. This is in contrast to the wobble-pairing of NNR codons with a single anticodon *UNN in prokaryotes. The difference between eukaryotes and prokaryotes is not due to the differences between the tRNAs, as anticodon *UNN of eukaryotic tRNAs recognizes both NNA and NNG codons when *E. coli* ribosomes are used. Rather, eukaryotes have larger and more complex ribosomes (80 S) than prokaryotes (70 S), and this difference may result in differences in codon–anticodon pairing.

The presence of anticodon *UNN in a family box, although very rare, is not deleterious if anticodon GNN is present. In fact, tRNA with anticodon $cmnm^5UCC$ is present for Gly in *Bacillus subtilis* (Murao and Ishikura, 1978).

2.2.3 *I*

I at the first anticodon position occurs in eukaryotic family boxes, except for Gly, and in the Arg CGN family box in eubacteria, with the exclusion of GNN anticodons. In eukaryotes, the enzyme hypoxanthine ribosyltransferase is thought to replace adenosine with hypoxanthine riboside (inosine) in the anticodon first position (Elliott and Trewyn, 1984). To make these replacements possible, it is thought that the ANN anticodon sequence in DNA may have been formed by mutation of GNN. The corresponding enzyme has not been reported for the bacterial Arg CGN family box. At any rate, the conversion of A to I seems to be specific to ACG to ICG in bacteria, because, as noted below, $tRNA^{Thr}_{AGU}$ occurs in *Mycoplasma* spp., and yet the A at the first anticodon position is unmodified (Andachi *et al.*, 1987, 1989; Samuelsson *et al.*, 1987).

In eubacteria, I recognizes U, C, and A of the third nucleoside of the codon, as originally proposed by Crick (1966*a*). This was confirmed by trinucleotide-dependent aminoacyl-tRNA binding to the ribosomes in *E. coli* system (Söll *et al.*, 1966*a,b*; Caskey *et al.*, 1968), and by a cell-free translation system of *Mycoplasma capricolum* using synthetic messenger RNAs containing in-frame CGN Arg codons, in which three codons—CGU, CGC, and CGA—are translated (Oba *et al.*, 1991*a*).

In eukaryotes, however, the decoding properties of I were thought to be different, as in the case of *U. *Schizosaccharomyces pombe* has at least two species of $tRNA^{Ser}$ for the UCN Ser family box, the major having anticodon IGA and the minor having *UGA [*U is partly 2-thiouridine and partly 5-(methylcarbonylmethyl) uridine]; it is likely that $tRNA^{Ser}_{CGA}$ also exists. The two species of $tRNAs^{Ser}_{*UGA}$ recognize codon UCA; haploid spores in which both of their genes have been genetically converted to stop-codon suppressors are lethal (Munz *et al.*, 1981). Munz *et al.* concluded that the function of $tRNA^{Ser}_{*UGA}$ in translating codon UCA cannot be taken over by any other $tRNAs^{Ser}$, including $tRNA^{Ser}_{IGA}$, i.e. inosine (I) cannot decode A *in vivo*. Although early work with trinucleotide-dependent eukaryotic aminoacyl-tRNA binding to ribosomes clearly demonstrated the wobble pairing of I with U, C, and A, the ribosomes used were again from *E. coli*, as was the case with *U. Guthrie and Abelson (1982) proposed the wobble rules for third position codon–anticodon pairing in yeast tRNAs in which *U pairs only with A, and I pairs only with U and C. However, Suzuki *et al.* (1994), using the *in vitro* translation system of *Candida cylindraceae*, demonstrated that the codon CUA was translated efficiently as Leu by $tRNA^{Leu}_{IAG}$. This indicates that, even in eukaryotes, I at the first wobble position of anticodon can base-pair with A at the third position of codon, at least *in vitro*. Thus, INN anticodons would produce intolerable ambiguity in pairs of 2-codon sets by simultaneously recognizing codons for different amino acids. Hence, the eukaryotic and prokaryotic codes contain INN anticodons only in family boxes, and never in 2-codon sets.

2.2.4 *A*

It had been believed that A at the first anticodon position was always modified to I (Lewin, 1990). It has now become evident that this rule is not absolute, because an unmodified A at the first anticodon position has been reported for yeast mitochondrial $tRNA^{Arg}_{ACG}$ (Sibler *et al.*, 1986) and for $tRNA^{Thr}_{AGU}$ of *Mycoplasma* spp. (Andachi *et al.*, 1987, 1989; Samuelsson *et al.*, 1987). *Aspergillus nidulans* mitochondrial $tRNA^{Gly}_{ACC}$ (Köchel *et al.*, 1981) and chloroplast $tRNA^{Arg}_{ACG}$ (determined by DNA sequencing) (Shinozaki *et al.*, 1986) might also have an unmodified A.

In *Mycoplasma* spp., $tRNA^{Thr}_{AGU}$ occurs with $tRNA^{Thr}_{UGU}$ in the Thr family box. The $tRNA^{Thr}_{AGU}$ gene has resulted from a mutation in the anticodon of the $tRNA^{Thr}_{UGU}$ gene, and not from the gene for $tRNA^{Thr}_{GGU}$. The anticodon AGU in *Mycoplasma capricolum* translates codons ACU, ACC, and ACG and, to a much lesser extent, AGA, *in vitro* (Kojima *et al.*, unpublished data; cited in Osawa *et al.*, 1992) (Fig. 2.4). $tRNA^{Gly}_{ACC}$ obtained by site-directed mutagenesis of the gene for *E. coli* $tRNA^{Gly}_{CCC}$ was found to read all four Gly codons *in vitro* (Boren *et al.*, 1993). Thus A, like unmodified U, seems to recognize all four bases.

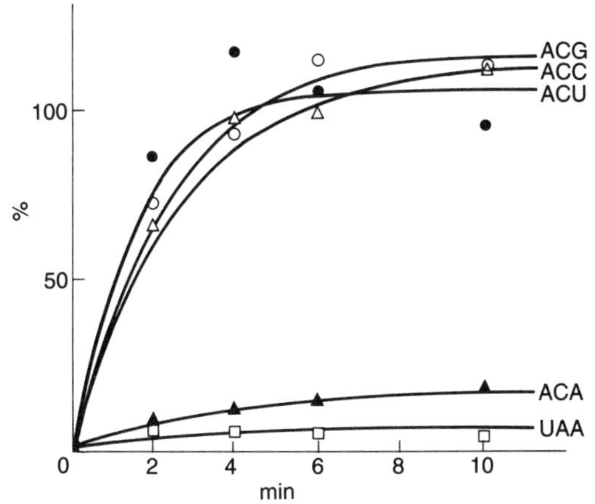

Fig. 2.4. Anticodon AGU reads all four codons in the Thr family box. Incorporation of [³H]radioactivity from [³H]threonyl-tRNA$^{\mathrm{Thr}}_{\mathrm{AGU}}$ into the peptide fraction was measured with the S30 fraction of *Mycoplasma capricolum*. For the mRNA used, see Fig. 2.1. (From Kojima *et al.*, unpublished.)

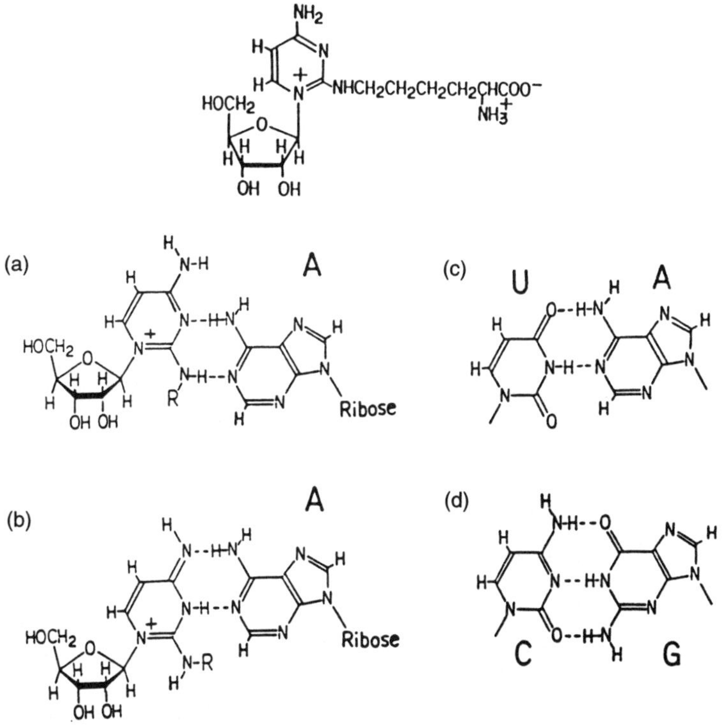

Fig. 2.5. Structure of lysidine and its pairing with adenosine. (Top) Structure of lysidine; (Bottom) base pairs (a and b) L·A; (c) U·A; and (d) C·G. R = Lys in (a) and (b) (from Muramatsu *et al.*, 1988*a*).

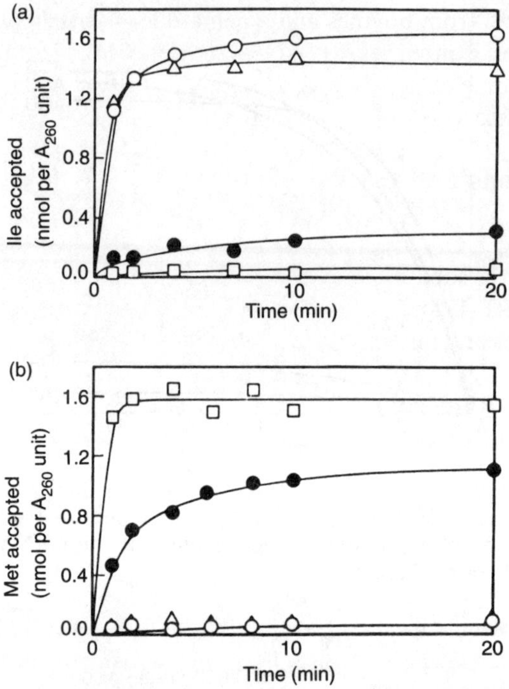

Fig. 2.6. (a) Isoleucylation and (b) methionylation of (○) $tRNA^{Ile}_{LAU}$; (●) $tRNA^{Ile}_{CAU}$; (△) $tRNA^{Ile}_{GAU}$; and (□) $tRNA^{Metm}_{ac^4CAU}$. Replacement of L(34) with unmodified C(34) (●) results in a marked reduction of the Ile-accepting activity (a), and in appearance of the Met-accepting activity (b). Measured with *E. coli* synthetase (from Muramatsu *et al.*, 1988*b*).

2.2.5 *C*

C at the first anticodon nucleoside pairs with G when it is unmodified, in accordance with the original proposal by Crick (1966*a*). The Ile codon AUA in eubacteria and plant mitochondria pairs with anticodon *CAU only when *C is 2-lysylcytidine, lysidine (Muramatsu *et al.*, 1988*a*,*b*; Weber *et al.*, 1990) (Fig. 2.5). The *E. coli* $tRNA^{Ile}_{LAU}$ is charged exclusively with Ile by isoleucyl-tRNA synthetase and recognizes only codon AUA. When L (lysidine) is replaced with unmodified C, the identity of this tRNA changes from Ile to Met, and the tRNA translates only codon AUG as Met (Muramatsu *et al.*, 1988*a*) (Fig. 2.6). $tRNA^{Ile}_{LAU}$ is the only example to have anticodon LNN.

Codon AUA is used for Met in mitochondria of yeast and most metazoans (see Section 6.3.2), and yet the anticodon of all $tRNA^{Met}$ genes is CAT. HsuChen *et al.* (1983) suggested that, in mitochondria, unmodified C in the wobble position can wobble-pair with A and G in the first codon nucleoside. This is a departure from the original wobble rules, and would mean that anticodon CAU translates AUG and AUA as Met. However, it has become evident that in mitochondrial $tRNA^{Met}$ from bovines and a nematode—*Ascaris suum*—C at the first anticodon position is modified to 5-formylcytidine (f^5C) (Moriya *et al.*, 1994). It is then possible for f^5C to pair with G and A.

2.3 Stop codons and release factors

UAA, UAG, and UGA are stop (termination) codons in the universal genetic code. In eubacteria such as *E. coli* and *Bacillus subtilis*, two protein factors—release factor 1 (RF-1) and release factor 2 (RF-2)—participate in the recognition of the stop codons and in the release of nascent polypeptides from the ribosomes. RF-1 recognizes UAA and UAG, and RF-2 recognizes UAA and UGA. As neither RF-1 nor RF-2 contains RNA, the recognition of a trinucleotide codon with respect to both base composition and amino acid sequence by RF protein molecules is an interaction of extremely high fidelity (Caskey *et al.*, 1969; Capecchi and Klein, 1969; Caskey, 1980). An alternative view would be that RFs possess a specific tertiary structure that can bind a tRNA-like molecule lacking an amino acid-binding site. This is analogous to the interaction between an aminoacyl-tRNA synthetase with the corresponding tRNA molecule, although such an RNA molecule has not been identified.

Göringer *et al.* (1991) proposed that, as the initial step of peptide chain termination, a direct interaction occurs between the UGA stop codon and the UCA motif (UCAUCA) within the 3′ major domain of *E. coli* 16S rRNA. This proposal is based on the observation that changing of the UCA sequences by site-directed mutagenesis disrupts base pairing with the UGA stop codon, but not with UAA or UAG, so permitting read-through of UGA. In *Mycoplasma capricolum*, however, the sequence CUACUA replaces UCAUCA, and yet RF from *E. coli* or *B. subtilis* recognizes the codon UGA (Inagaki *et al.*, 1993) (see Section 6.5.1). Even if the 16S rRNA-stop codon interaction model is correct, the question of how RF can recognize UGA remains unsolved. Furthermore, the sequence motifs that might be responsible for the other two stop codons have not been recognized.

2.4 Four-way wobbling versus the 'two-out-of-three' mechanism

As noted above, a single anticodon—UNN (U unmodified)—exists in most or all family boxes of non-plant mitochondria and *Mycoplasma* spp. The questions then arise: (1) is the U in the first anticodon position specific for family box pairs with all four bases? or (2) are these codons translated by the 'two-out-of-three' mechanism without discrimination between the nucleosides in the third codon position and the first anticodon position, so that only the second and third positions of the anticodon pair with the codon?

In his two-out-of-three hypothesis, Lagerkvist (1981) states that:

Codons may be read by relying mainly on the Watson–Crick base pairs formed with the first two codon positions, while the mispaired nucleotides in the third codon and anticodon wobble positions make only a marginal contribution to the total stability on the reading interaction.

He further notes that:

The probability of such reading would be greatest for codons making only G–C interactions in these (the first and second) positions ('strong codons'), intermediate for codons which make one A–U and one G–C interaction ('mixed codons') and minimal for codons making only A–U interactions ('weak codons') . . . strong codons are without exception restricted to the codon families, while weak codons are always outside the families.

This is an attractive hypothesis, and the following findings may be explained by this mechanism. In yeast mitochondria, only one species of tRNA—$tRNA^{Arg}_{ACC}$—exists for CGN codons (strong codons) and yet codon CGG is used. The same is true for tobacco and rice chloroplasts, although liverwort chloroplasts have both $tRNA^{Arg}_{AGG}$ (modification unknown) and $tRNA^{Arg}_{CCG}$. Sibler *et al.* (1986) and Shinozaki *et al.* (1986) state that, like anticodon UCG, anticodon ACG (or possibly ICG in chloroplasts) would translate all the CGN codons by a two-out-of-three mechanism, because A (anticodon) was thought to pair only with U (codon).

However, in a refined cell-free translation system of *Mycoplasma capricolum*, anticodon ICG (Arg) translates codons CGU, CGC, and CGA according to the standard wobble rules, and yet codon CGG is not translated at all (Oba *et al.*, 1991*a*). The two-out-of-three mechanism does not occur even for the 'strong codons'. Furthermore, it is now apparent that all *Mycoplasma* tRNAs with unmodified U or A at anticodon first position can read the corresponding four codons in family boxes with different efficiencies *in vitro* (Inagaki *et al.*, unpublished). It is thus more likely that the first nucleoside U or A of the anticodon pairs, by four-way wobble, with U, C, A, or G. Nucleotide C32 or modification of R37 is important for an appropriate conformation of tRNA for this wobbling (Lustig *et al.*, 1993; Inagaki *et al.*, unpublished).

2.5 Role of nucleotide 37 in codon–anticodon pairing

Codon–anticodon pairings seem to be affected by some other part of tRNA molecules as well as the anticodon. The most important is the nucleotide 37, next to the 3′-side of the anticodon, which is usually extensively modified. The modifications may be concerned with strengthening the weak U : A or A : U pairing between the third anticodon nucleotide (position 36) and the first codon nucleotide.

2.5.1 *U-started codons*

U-started codons must pair with A-ended anticodons (A3 : U1 pairing), because a mispairing of U at the first position of codons with G at the third anticodon position (G3 : U1 pairing) results in the misreading of the codons to different amino acids. The common modification at 37 of tRNAs for UNN codons usually has a bulky hydrophobic group, such as a large isopentenyl side-chain attached to adenine (N^6-isopentenyladenosine; i^6A) or its derivative (ms^2i^6A, ms^2io^6A), or the Y base in most of eukaryotic $tRNAs^{Phe}$ (Jukes, 1973*a*; Björk, 1987; Nishimura, 1979; Kano *et al.*, 1991). These bulky hydrophobic modifications are thought to play a role in strengthening the A^{36}–U interaction. However, this suggestion was questioned because of a lack of these bulky side-chains at 37 in some of the eukaryotic, eubacterial, and archaebacterial tRNAs reading UNN codons (see below and Table 2.3) (Björk *et al.*, 1987). The other modifications are 1-methylguanosine (m^1G) and 6-methyladenosine (m^6A). These are present mainly in tRNAs from *Mycoplasma* spp. (Andachi *et al.*, 1989). A37 is sometimes unmodified, such as in tRNA with anticodon CAA (Leu) in *Micrococcus luteus* or GGA (Ser) in both *M. luteus* (Kano *et al.*, 1991) and *E. coli* (Nishimura, 1979). More studies are needed as to the significance of A37 modifications in codon–anticodon interactions.

2.5.2 *A-started codons*

A-started codons must pair with U-ended anticodons (U3 : A1 pairing) to avoid U3 : G1 mispairing. The modification of A37 in tRNAs for ANN codons is usually a bulky threonyl-containing side-chain at position 6 of adenine [*N*-((9-β-D-ribofuranosyl purine-6-yl)carbamoyl)threonine] (t^6A) (Nishimura, 1979; Björk, 1987). It was proposed that steric hindrance by such a bulky side-chain prevents wobble-pairing of U3 : G1 (Jukes, 1973*a*). There are two exceptions to this. Initiator $tRNA^{Met\text{-}i}$ from prokaryotes, including chloroplasts, has unmodified A in all its examined sequences (Sprinzl *et al.*, 1989). It is significant that initiator methionine in prokaryotes is often encoded by GUG, and that unmodified A37 facilitates U3 : G1 pairing (Jukes, 1973*a*). A37 of elongator $tRNA^{Met\text{-}m}_{CAU}$ is not modified in *Mycoplasma capricolum*, whereas A37 of this tRNA species in other organisms is always modified to t^6A or its derivative, as is the case with other tRNAs for ANN codons (Andachi *et al.*, 1989). However, codon GUG (Val) is used rarely (0.2 per cent among all codons) in this bacterium, so that an occasional U3 : G1 mispairing between anticodon CAU (Met) and codon GUG (Val) (if it occurs at all), may not be serious.

2.5.3 *C-started codons*

C-started codons, e.g. Leu (CUN), Pro, His, Gln, and Arg (CGN), are read by anticodons ending with G. Most nucleotides in position 37 in tRNAs for these amino acids have a simple modification, e.g. 1-methyl-G (m^1G) or 2-methyl-A (m^2A) (Jukes, 1973*a*; Nishimura, 1979; Björk *et al.*, 1987; Kano *et al.*, 1991). The more stable G3 : C1 pairs may be enough to stabilize the codon–anticodon interactions than the less stable A3 : U1 pairs, so that a bulky side chain at 37 of tRNAs may not be required to prevent mispairing.

Table 2.2 Known anticodons with nucleosides at position 37 in parentheses

Anticodon	L	E	B	M
Phe (UUU)	GAA(N2)	GAA(ms^{2}i^{6}A)	GAA(ms^{2}i^{6}A)	GAA(m^{1}G)
Phe (UUC)				
Leu (UUA)		UAA(ms^{2}i^{6}A)	UAA	UAA(m^{6}A)
Leu (UUG)	CAA(A)	CAA	CAA	CAA(m^{6}A)
Leu (CUU)	GAG(m^{1}G)	GAG(m^{1}G)		UAG(m^{1}G)
Leu (CUC)				
Leu (CUA)		UAG		
Leu (CUG)	CAG(m^{1}G)	CAG(m^{1}G)	CAG(m^{1}G)	
Ile (AUU)	GAU(N1)	GAU(t^{6}A)	GAU	GAU(t^{6}A)
Ile (AUC)				
Ile (AUA)		LAU(t^{6}A)	CAU	LAU(m^{6}A)
fMet (AUG)	CAU(A)	CAU(A)	CAU	CAU(A)
Met (AUG)	CAU	CAU(t^{6}A)	CAU(m^{2}A,m^{6}A)	CAU(A)
Val (GUU)	GAC(A)	GAC(A)		UAC(m^{6}A)
Val (GUC)				
Val (GUA)		UAC(m^{6}A)	UAC(ms^{2}i^{6}A)	
Val (GUG)	CAC(A)			
Tyr (UAU)	GUA§	GUA(ms^{2}i^{6}A)	GUA(ms^{2}i^{6}A,i^{6}A)	GUA(m^{6}A)
Tyr (UAC)				
– (UAA)				
– (UAG)				
His (CAU)	GUG(m^{1}G)	GUG(m^{2}A)	GUG(m^{1}G)	GUG(m^{1}G)
His (CAC)				
Gln (CAA)		UUG(m^{2}A)	UUG	UUG(m^{6}A)
Gln (CAG)	CUG(m^{1}G)	CUG(m^{2}A)		
Asn (AAU)	GUU(N1)	GUU(t^{6}A)		GUU(t^{6}A)
Asn (AAC)				
Lys (AAA)		UUU(t^{6}A)	UUU(ms^{2}t^{6}A)	UUU(t^{6}A)
Lys (AAG)	CUU(t^{6}A)			CUU(t^{6}A)
Asp (GAU)	GUC(A)	GUC(m^{2}A)	GUC	GUC(A)
Asp (GAC)				
Glu (GAA)		UUC(m^{2}A)	UUC	UUC(A)
Glu (GAG)	CUC§			

in several eubacteria (footnotes on following page)

Anticodon	L	E	B	M
Ser (UCU)	GGA(A)	GGA(A)	GGA	UGA(m^{6}A)
Ser (UCC)				
Ser (UCA)		UGA(ms^{2}i^{6}A)	UGA	
Ser (UCG)	NGA(i^{6}A)†	CGA(ms^{2}i^{6}A)		
Pro (CCU)	GGG(m^{1}G)	GGG		UGG(m^{1}G)
Pro (CCC)				
Pro (CCA)	UGG	UGG	UGG(m^{1}G)	
Pro (CCG)	CGG(m^{1}G)	CGG		
Thr (ACU)	GGU(t^{6}A)	GGU(mt^{6}A)	GGU	AGU(t^{6}A)
Thr (ACC)				
Thr (ACA)		UGU(t^{6}A)	UGU(t^{6}A)	UGU(t^{6}A)
Thr (ACG)	CGU(t^{6}A)	CGU		
Ala (GCU)	GGC(A)	GGC(A)		UGC(m^{6}A)
Ala (GCC)				
Ala (GCA)		UGC(A)	UGC(m^{6}A)	
Ala (GCG)	CGC(A)			
Cys (UGU)	GCA(i^{6}A)	GCA(ms^{2}i^{6}A)	GCA	GCA(m^{6}A)
Cys (UGC)				
– (UGA)				UCA(m^{6}A)
Trp (UGG)	CCA(i^{6}A)	CCA(ms^{2}i^{6}A)	CCA	CCA(m^{6}A)
Arg (CGU)	ICG(m^{1}G)	ICG(m^{2}A)	ICG(m^{1}G)	ICG(m^{1}G)
Arg (CGC)				
Arg (CGA)				
Arg (CGG)	CCG(m^{1}G)	CCG(m^{1}G)		–
Ser (AGU)		GCU(t^{6}A)	GCU	GCU(t^{6}A)
Ser (AGC)				
Arg (AGA)		UCU(t^{6}A)		UCU(t^{6}A)
Arg (AGG)		CCU(t^{6}A)		
Gly (GGU)	GCC(A)	GCC(A)	GCC	UCC(m^{6}A)
Gly (GGC)				
Gly (GGA)		UCC(A)	UCC(A)	
Gly (GGG)	CCC(A)	CCC(A)		

Table 2.3 Nucleosides at position 37 in tRNA from eukaryotes, eubacteria, and archaebacteria

Codon (nucleotide)			Nucleosides at position 37 in tRNA		
1st	2nd	3rd	Eukaryotes	Eubacteria	Archaebacteria (Metabacteria)
U	N	N	yW; o^2yW; i^6A; m^1G	A; m^6A; ms^2i^6A; i^6A; ms^2io^6A; m^1G	m^1G
C	N	N	m^1G	m^1G; m^2A	m^1G
A	N	N	t^6A; mt^6A; ms^2t^6A	A; t^6A; mt^6A; m^6A; ms^2t^6A	t^6A
G	N	N	m^1G; m^1I	A; m^2A; m^6A; ms^2i^6A	m^1G

yW: nucleoside of Y base; o^2yW: nucleoside of peroxy Y base. For abbreviations of other nucleosides, see legend for Table 2.2. From Björk *et al.* (1987), with additions and modifications. N : A, G, U, or C.

Abbreviations at position 37 nucleoside: m^2A, 2-methyladenosine; ms^2i^6A, 2-methylthio-N^6-isopentenyladenosine; m^6A, N^6-methyladenosine; t^6A, *N*-((9-β-D-ribofuranosylpurine-6-yl)carbamoyl)-threonine; mt^6A, *N*-((9-β-D-ribofuranosylpurine-6-yl)*N*-methyl-carbamoyl)-threonine; m^1G, 1-methylguanosine; N1 and N2, unidentified. Modifications of the anticodon 1st nucleoside are not indicated except for L, 2-lysylcytidine (Ile) and I, inosine (Arg). Anticodons not showing position 37 nucleosides are from DNA sequences.

L, *Micrococcus luteus*. Distinction between i^6A and ms^2i^6A is not made, and it is designated i^6A in this table. †, Unidentified; §, tentative. E, *Escherichia coli*; B, *Bacillus subtilis*; M, *Mycoplasma capricolum*. —, most probably the unassigned codon (from Kano *et al.*, 1991).

2.5.4 *G-started codons*

G-started codons interact with C-ended anticodons. As is the case with C-started codons, C3 : G1 pairing is strong enough to prevent mispairing, so that modification of A37 would not be so essential. In fact, position 37 in these tRNAs has a simple modification, such as m^6A, ms^2i^6A, or m^2A in *E. coli*, *B. subtilis*, and *M. capricolum* (Nishimura, 1979; Andachi *et al.*, 1989). However, it is always unmodified A in *Micrococcus luteus* ($tRNA^{Val}_{GAC}$, $tRNA^{Val}_{CAC}$, $tRNA^{Ala}_{GGC}$, $tRNA^{Ala}_{CGC}$, $tRNA^{Asp}_{GUC}$, $tRNA^{Glu}_{CUC}$, $tRNA^{Gly}_{GCC}$, and $tRNA^{Gly}_{CCC}$) (Kano *et al.*, 1991).

Modification profiles of position 37 have considerable regularities (Tables 2.2 and 2.3), although this is not consistent between different organisms or even between similar codon–anticodon interactions. It would be interesting to discover whether these variations reflect fine adjustments of codon–anticodon pairing in various cases in translation.

2.6 tRNA identity

2.6.1 *Organisms*

The aminoacylation of tRNA is catalysed by aminoacyl-tRNA synthetase (ARS), and results in the joining of the amino acid to the 3′-end of adenosine in the tRNA molecule. There is one species of synthetase for each amino acid. Each synthetase aminoacylates all of the isoacceptor tRNAs, proving that the synthetases are amino acid-specific. The correct translation of a codon is mediated by the anticodon of the aminoacyl-tRNA. The prerequisite of this process is the correct attachment of amino acids to the cognate tRNAs by ARS. Each ARS must recognize its cognate tRNA from the other tRNA species, all of which possess very similar secondary and tertiary structures. The characteristic element of each tRNA that is recognized by its cognate ARS is the tRNA identity element (Schimmel, 1987; Schulman and Abelson, 1988; Yarus, 1988; Normanly and Abelson, 1989; McClain, 1993; Schimmel *et al.*, 1993). Experiments to establish the tRNA identity element have been carried out by using *in vitro* transcripts of tRNA into which base replacements have been introduced at various positions. Another approach is the *in vivo* suppressor assay, which examines whether tRNA genes with their anticodon replaced by suppressor CUA or UCA anticodon sequence, derived either genetically or by synthetic means, were expressed and charged by the amino acid corresponding to the original tRNA species (Normanly and Abelson, 1989; McClain, 1993). In the main, these experiments have been carried out on *E. coli*, although yeast systems have also been used. Generally speaking, major identity determinants are localized in the anticodon region and in the acceptor stem, including the discriminator base (Fig. 2.7), although in some cases ($tRNA^{Ala}$, $tRNA^{Leu}$, and $tRNA^{Ser}$) no identity elements have been detected in the anticodon region.

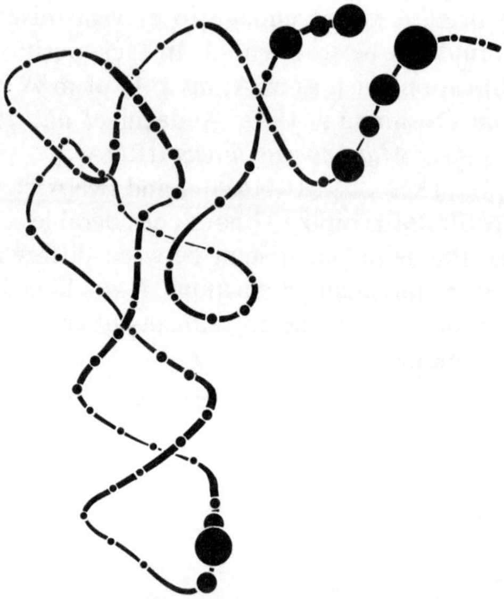

Fig. 2.7. Distribution of tRNA identity determinant nucleotides in *E. coli.* A circle diameter is proportional to the fraction of the 20 tRNA acceptor types (from McClain, 1993).

The crystal structure of *E. coli* $tRNA^{Gln}$-synthetase/$tRNA^{Gln}$ complex (Rould *et al.*, 1989), and of yeast Asn-synthetase/$tRNA^{Asn}$ complex (see Cavarelli and Moras, 1993, for a review) shows a direct interaction of each synthetase with the acceptor helix and anticodon, together with the presence of a specific amino acid binding site on the ARS. This suggests that both the acceptor helix and anticodon are involved in the specific binding of amino acids to the 3′-end of the tRNA molecule. The results agree perfectly with the tRNA identity profile determined by the base-replacement of tRNA. Nureki *et al.* (1994) demonstrated that, in *E. coli* $tRNA^{Ile}_{GAU}$ and $tRNA^{Ile}_{LAU}$, the identity elements necessary for isoleucylation through chemical interaction with IleRS exist in the acceptor, D-, and anticodon stems, and in the anticodon loop. Interestingly, two tertiary base pairs near the identity elements are indispensable for isoleucylation. Nureki *et al.* propose that: 'the recognition by IleRS of all the widely distributed identity determinants is coupled with a global conformational change that involves the loosening of a particular set of tertiary base-pairs of $tRNA^{Ile}$'.

Note that the identity elements of all the *E. coli* tRNA species are in the acceptor stem, although the anticodon is involved in identity in some tRNAs.

This would indicate that some ARSs, such as for Ala and Ser, do not interact with anticodon. Instead they react with the acceptor helix, suggesting that the primary importance of the tRNA identity resides in the acceptor helix, although the anticodon region plays a role in other tRNAs (Hou and Schimmel, 1988; see Schimmel *et al.*, 1993 for review). Conversely, it may also be argued that the anticodon is of primary importance because it recognizes only codons for a single amino acid. The apparent unimportance of the anticodon sequence in some tRNA species (such as $tRNA^{Ser}$) for recognition by ARS is the conseqeunce of evolution. It increases the role of other regions as the identity elements of the tRNA (Asahara *et al.*, 1994).

Schimmel *et al.* (1993) noted that eight amino acids—Ile, Met, Val, Ala, Asn, Gly, His, and Ser—were charged with ARS to the specific RNA substrates of acceptor helix domains of various sizes (all of which lack distal parts of the tRNA structure, including anticodon). The aminoacylation was found to be sequence-specific. Generally, 'N37 discriminator base and certain base pairs within the first four of the acceptor stem are needed to confer aminoacylation and to determine specificity' (Schimmel *et al.*, 1993). For example, charging of Ala with alanylsynthease is dependent on the C3 : U70 base pair and that of His on the unique extra base pair C1 : C7.

Twenty ARS species may be structurally classified into two groups (Cavarelli and Moras, 1993). The class I enzymes are mostly monomeric and include ARS for Arg, Cys, Glu, Gln, Ile, Leu, Met, Tyr, Trp, and Val. The class II enzymes, most of which are multimeric homopolymers, include ARS for Ala, Asn, Asp, Gly, His, Lys, Phe, Pro, Ser, and Thr. The class I and II enzymes have no homology with respect not only to primary structure but to tertiary structures as well, suggesting that the origins of the two classes are independent. The ARSs are composed of two domains; (1) the conserved domain, including the regions with sequence similarities characteristic for each class as part of the active site structure; and (2) the non-conserved domain, showing little sequence similarity in either class.

The conserved active domain incorporates the determinants for acceptor helix interactions, and the non-conserved domain interacts with distal parts of tRNA that include the anticodon. As the non-conserved domain contributes little to the recognition of the acceptor helix, the conserved domain 'behaves as though it is operationally independent of the non-conserved domain'. Schimmel *et al.* (1993) defined an operational RNA code as 'the specific aminoacylation reactions with RNA substrates that lack the anticodons of the classical genetic code', which is based on the conserved domain of synthetases. 'More narrowly defined . . . the 7 bp acceptor helix terminating in the single stranded NCCA-3′ OH can be considered' (Schimmel *et al.*, 1993).

However, there is no doubt that in some cases (e.g. *E. coli* $tRNA^{Ile}$), anticodon is a strong identity determinant. A remarkable example is that the *E. coli* $tRNA^{Ile}_{LAU}$ is charged exclusively with Ile by isoleucyl-tRNA synthetase (IleRS), and recognizes only the AUA codon. When L is replaced by unmodified C, the tRNA is no longer recognized by IleRS. This tRNA is charged with Met by

MetRS, and translates only codon AUG as Met (Muramatsu *et al.*, 1988*a*) (see Fig. 2.6). These results clearly show that modification from C to L of the first anticodon nucleoside is responsible both for the tRNA identity and for recognition of the third nucleoside A of codon AUA. The result does not necessarily mean that the acceptor stem is not involved, as both $tRNA^{Met}$ and $tRNA^{Ile}$ have other identity elements in this and other regions. AUA was supposed to be a Met codon in the early code and is in fact so in many mitochondria. The $tRNA^{Ile}_{LAU}$ could have originated relatively recently from $tRNA^{Met}_{CUA}$, and its structure could have evolved so as to enable it to interact with IleRS along with modification of C to L at the anticodon first position. Another example is that substitution of U35 with C35, or U36 with C36 (anticodon base) abolishes aminoacylation of *E. coli* $tRNA^{Asn}$. Substitution of G34 with C34, and G73 with A73 (discriminator base of $tRNA^{Lys}$) converts $tRNA^{Asn}$ into a Lys acceptor. These experiments show that both the anticodon and discriminator base are important for aminoacylation of $tRNA^{Asn}$ (Li *et al.*, 1993). The abolition of aminoacylation by mutation of the anticodon base is apparently in contradiction to the result described by Schimmel *et al.* (1993), in that Asn is charged with AsnRS to the acceptor helix domain lacking anticodon. The binding of AsnRS to a false anticodon might weaken its affinity to the acceptor stem of tRNA. As for the partial discrepancy between the identified determinants and the specific aminoacylation of minihelices as exemplified above, McClain (1993) noted that the experiments with minihelices 'provide an incomplete picture of the cellular process', but 'suggest that similar small RNAs preceded tRNAs in evolution.' For a comprehensive and excellent review article on the tRNA identity, see McClain (1993).

2.6.2 *Mitochondria*

Bovine mitochondrial ARSs (Phe, Thr, Arg, and Lys) can aminoacylate their cognate tRNAs from *E. coli*, *Thermus thermophilus*, and mitochondria, although none of the *E. coli* enzymes can do mitochondrial tRNAs ('unilateral aminoacylation') (Kumazawa *et al.*, 1989, 1991). Thus, the tRNA recognition mechanisms by ARS could be conserved, at least in part, between bacteria and mitochondria. Kumazawa *et al.* (1991) assume that the identity determinants in the majority of mitochondrial tRNAs may be centred on the anticodon sequences, because anticodons are conserved sequences only between bacteria and mitochondria. The authors further suggest that the unilaterality could be the result of co-evolutionary simplification of ARS, with retrogression of mitochondrial tRNA structure. Whether identity elements are localized only in anticodons and not in the acceptor stem needs direct demonstration of the elements, because the sequence of the acceptor stem could be changed with change(s) of ARS structure, or conformation of this region could be conserved even if the sequence underwent some changes. The major identity elements of mitochondrial $tRNA^{Ser}$ do not seem to include the anticodon.

During evolution, identity elements could have changed, at least to some extent, in response to structural changes of tRNA. These changes have in turn affected the recognition mechanism of ARSs, as seen in mitochondria. It is important to survey identity elements of tRNA in a wider range of organisms and organelles to establish a unified view of tRNA identity.

Change in the tRNA identity naturally causes change in the meaning of codon(s) pairing with the tRNA. The identity change may be involved in amino acid reassignment of a codon, such as assumed in CUN Leu to Thr or AUA Ile to Met in mitochondria (see Sections 6.6.2 and 6.6.3), or AUA Met to Ile in early evolution of the code (see Section 9.3.4).

3 Anticodon composition

The number of anticodons is always less than the total number of amino acid codons (of which there are 61), although the precise number varies; the smallest is 22 (in vertebrate mitochondria) and the largest possible is 46, although no organism has this many. The recognition pattern of a synonymous codon set by tRNA shows considerable variation between organisms. The anticodon composition lists for various organisms and organelles are given in Tables 3.1 to 3.5. The variations seen result sometimes from codon usage and sometimes from the tendency to reduce the genome size (genomic economization). In some cases, certain codons and the corresponding tRNAs disappear from the genome (see Sections 5.1 and 5.2).

3.1 Bacteria

Table 3.1 shows the anticodon species of several eubacteria and fungal mitochondria. A higher genomic G + C content is accompanied by an increase in G or C in silent codon positions, and as CNN anticodons translate NNG codons, one would expect more CNN anticodons in eubacteria with a higher G + C content; this is the case ('CNN rule'). *Micrococcus luteus* (G + C content = 75 per cent) has 13 CNN anticodons (Kano *et al.*, 1991), and two more (CGA and CCU) are presumed to exist. *E. coli* (G + C content = 50 per cent) has 11 CNN anticodons (Komine *et al.*, 1990), and *Thermus thermophilus* (G + C content = 69 per cent) has three more anticodons (CAC, C̣UU, and CUC; Hara-Yokoyama *et al.*, 1986) that are not found in *E. coli*. *B. subtilis* (G + C content = 43 per cent) has at least six CNN anticodons (Vold, 1985), and *Mycoplasma capricolum* (G + C content = 25 per cent) has only five (Andachi *et al.*, 1989). In *Micrococcus luteus*, UNN (modified U) and LAU (L is lysidine) anticodons have become a trace or non-existent (Kano *et al.*, 1991).

The eight family boxes in *Micrococcus luteus* (Kano *et al.*, 1991), *E. coli* (Komine *et al.*, 1990), and many other bacteria have seven GNN anticodons. These are absent from *Mycoplasma capricolum* and a single UNN anticodon (unmodified U), which pairs with four codons, is present in six of the eight family boxes (Andachi *et al.*, 1989). The higher the G + C content of the genome, the higher the CNN anticodon content and the lower the UNN anticodon content. In contrast, the lower the G + C content, the lower the CNN and GNN anticodon contents.

The anticodon composition of *M. pneumoniae* is somewhat different from that of *M. capricolum* (Simoneau *et al.*, 1993). In *M. pneumoniae*, CGG, which is absent in *M. capricolum*, is assigned to Arg and translated by $tRNA^{Arg}_{UCG}$, and CNN or GNN anticodons are found in the Ser, Thr, Arg, and Gly family boxes. As the genomic G+C content of *M. pneumoniae* is 40 per cent, which is much higher than that of *M. capricolum*, AT-pressure exerted in the *M. capricolum* lineage would have been switched to GC-pressure in the *M. pneumoniae* line, resulting in the appearance of CNN and GNN anticodons and in the abolition of preference towards AT-rich synonymous codons (see Section 4.2).

The anticodon compositions described above are the consequences of evolution, but are not tightly linked to the specific phylogenetic lines. For example, in the archaebacterial group, which is phylogenetically more related to eukaryotes than to eubacteria (Hori and Osawa, 1987; Iwabe *et al.*, 1989), anticodon usage follows the 'CNN rule'. Fourteen CNN anticodons have been reported from *Halobacterium volcanii* (G+C content = 63 per cent) but only one has been found in methanogens (G+C content = 30 per cent), although two obligatory CNN anticodons (CAU and CCA) are presumed to exist in methanogens (Osawa and Jukes, 1988; Osawa *et al.*, 1990*c*). According to Gupta (1984), no INN anticodons are present in *H. volcanii*. The halobacterial code is the same as the eubacterial code except for the absence of ICG (Arg) and the presence of GCG, UCG, and CCG (Table 3.2).

3.2 Chloroplasts and mitochondria

Chloroplasts contain DNA that is rich in A and T (28 to 39 per cent G+C in DNA). They follow the same rule as eubacteria; there are only five CNN anticodons, and GNN anticodons are absent from two family boxes. The chloroplast code is identical to the universal code. Two family boxes use single UNN anticodons (Shinozaki *et al.*, 1986; Umesono and Ozeki, 1987) (Table 3.3).

Most mitochondria, except for those in green plants, have one or two CNN anticodons, one of which is CAU. This apparently translates both Met codons, AUA and AUG (see Section 6.3.2). GNN anticodons are absent from all eight family boxes, which use single UNN anticodons (Table 3.4; see Osawa *et al.*, 1992, for a review).

3.3 Resemblance between *Mycoplasma* spp. and mitochondria

The *Mycoplasma* genomes are the smallest for all known free-living organisms (Muto *et al.*, 1991). They are regarded as degenerate forms of Gram-positive bacteria.

The *Mycoplasma capricolum* genome contains 30 tRNA genes for 29 tRNA species (Muto *et al.*, 1990). This is a much smaller number than in the genome of *E. coli*, which has 78 genes for 45 tRNA species (or 41 anticodon species) (Komine *et al.*, 1990), or *B. subtilis*, with at least 51 genes for 31 different tRNA species (Vold, 1985). Only $tRNA^{Lys}_{UUU}$ in *Mycoplasma capricolum* is encoded by

Table 3.1 Anticodons in eubacteria and fungal mitochondria [a,b] (footnotes on following page)

Amino acid (codon)	Anticodon in L	E	M	mt	Amino acid (codon)	Anticodon in L	E	M	mt
Phe (UUU)	GAA	GAA	GAA	GAA	Ser (UCU)	GGA	GGA		
Phe (UUC)					Ser (UCC)			UGA	UGA
Leu (UUA)		[1]UAA	[1]UAA	[2]UAA	Ser (UCA)		[7]UGA		
Leu (UUG)	CAA	[3]CAA	[3]CAA		Ser (UCG)	NGA	CGA		
Leu (CUU)	GAG	GAG			Pro (CCU)	GGG	GGG		
Leu (CUC)			UAG	UAG	Pro (CCC)			UGG	UGG
Leu (CUA)	UAG	[4]UAG			Pro (CCA)		[4]UGG		
Leu (CUG)	CAG	CAG			Pro (CCG)	CGG	CGG		
Ile (AUU)	GAU	GAU	GAU	GAU	Thr (ACU)	GGU	GGU	AGU	
Ile (AUC)					Thr (ACC)			UGU	UGU
Ile (AUA)	–	[5]LAU	[5]LAU	CAU[f]	Thr (ACA)	UGU[e]	[4]UGU		
Met (AUG)	CAU	[6]CAU	CAU		Thr (ACG)	CGU	CGU		
Val (GUU)	GAC	GAC			Ala (GCU)	GGC	GGC		
Val (GUC)			UAC	UAC	Ala (GCC)			UGC	UGC
Val (GUA)		[7]UAC			Ala (GCA)		[7]UGC		
Val (GUG)	CAC				Ala (GCG)	CGC			

Amino acid (codon)	Anticodon in L	E	M	mt	Amino acid (codon)	Anticodon in L	E	M	mt
Tyr (UAU)	GUA	^{8}QUA	GUA	GUA	Cys (UGU)	GCA	GCA	GCA	GCA
Tyr (UAC)					Cys (UGC)				
Stop (UAA)					Stop (UGA)			^{1}UCA ^{3}CCA	^{2}UCA^{c}
Stop (UAG)					Trp (UGG)	^{3}CCA	^{3}CCA		
His (CAU)	GUG	^{8}QUG	GUG	GUG	Arg (CGU)				
His (CAC)					Arg (CGC)	^{11}ICG	^{11}ICG	^{11}ICG	U/ACG^{d}
Gln (CAA)		^{9}UUG CUG	^{2}UUG	^{4}UUG	Arg (CGA)				
Gln (CAG)	CUG				Arg (CGG)	CCG	CCG	–	
Asn (AAU)	GUU	^{8}QUU	GUU	GUU	Ser (AGU)	GCU	GCU	GCU	GCU
Asn (AAC)					Ser (AGC)				
Lys (AAA)		^{10}UUU	^{2}UUU CUU	^{2}UUU CUU^{g}	Arg (AGA)	–	^{12}UCU CCU	^{2}UCU	^{2}UCU
Lys (AAG)	CUU				Arg (AGG)	CCU			
Asp (GAU)	GUC	^{8}QUC	GUC	GUC	Gly (GGU)	GCC	GCC		ACC^{h}
Asp (GAC)					Gly (GGC)			UCC	UCC
Glu (GAA)		^{10}UUC	^{2}UUC	^{4}UUC	Gly (GGA)		^{13}UCC CCC		
Glu (GAG)	CUC				Gly (GGG)	CCC			

two tRNA genes with identical sequences. The *Mycoplasma* tRNAs are characterized by a low content of modified nucleosides (Andachi *et al.*, 1989). As a consequence of AT-pressure and genomic economization, the genomes of *Mycoplasma capricolum* and mitochondria seem to have discarded the genes for many redundant and non-obligate tRNAs, and also the genes for many enzymes for tRNA nucleoside modifications. This has resulted in a similarity of tRNA usage and the anticodon lists.

3.4 Eukaryotes

Forty-four anticodons have been discovered in eukaryotes, and one more—UAA—is presumed to exist (Table 3.5). The list of eukaryotic anticodons differs conspicuously from the lists of anticodons of other organisms. It contains eight anticodons with inosine (I) in the first position, and carries a full complement of 15 CNN anticodons (Osawa and Jukes, 1988; Sprinzl *et al.*, 1989; Osawa *et al.*, 1992). The gene for the tRNA species with the same anticodon quite often exists in multiple. For example, at least 118 tRNA genes exist in the yeast *Saccharomyces cerevisiae* (Guthrie and Abelson, 1982), and at least 68 in the cellular slime mould *Dictyostelium discoideum* (Hofmann *et al.*, 1991). An increase in copy number usually contributes to the abundance of the tRNA populations, so as to accommodate the translation of the highly used codon(s). For example, only a single gene for $tRNA^{Ser}_{CAG}$ translating CUG codon (see Section 6.5.4) has been detected in *Candida albicans*, in which CUG is a rare codon. However, $tRNA^{Ser}_{CAG}$ is a predominant Ser codon in *C. cylindracea*, in which several copies of the genes for $tRNA^{Ser}_{CAG}$ have been detected (Watanabe, K., personal communication). In the yeasts and the slime mould, abundant codons have multiple copies (sometimes more than ten copies) of their corresponding tRNA genes, although there are some exceptions.

---

[a] Reproduced from Andachi *et al.*, 1987 with modifications.

[b] Abbreviations: L, *Micrococcus luteus*; E, *E. coli*; M, *Mycoplasma capricolum*; mt, fungal mitochondria; –, probably unassigned codon (most of the anticodons were determined in *Micrococcus luteus* (Ikeda *et al.*, 1990*a*, 1990*b*; Kano *et al.*, 1991)). Some more may exist; the nucleoside and its modification at the first anticodon position N for $tRNA^{Ser}$ are unknown. In yeast (*S. cerevisiae* and *T. glabrata*) mitochondria, both AUA and AUG code for Met and CUN codes for Thr. In *T. glabrata* mitochondria, CGN are probable unassigned codons (Clark-Walker *et al.*, 1985; Osawa *et al.*, 1990*a*). *Mycoplasma capricolum* and yeast mitochondria, UGA codes for Trp. Superscript numbers represent modifications of anticodon first nucleosides as follows: 1, 5-carboxymethylaminomethy-2′-*O*-methyluridine ($cmnm^5Um$); 2,5-carboxymethylaminomethyluridine ($cmnm^5U$); 3, 2′-*O*-methylcytidine (Cm); 4, probably modified; 5, 4-amino-2-(N^6lysino-1-β-D-ribofuranosyl)pyrimidinium (L, lysidine); 6, N^4-acetylcytidine (ac^4C); 7, uridine-5-oxyacetic acid (o^5U); 8, queuosine (Q); 9, probably 2-thiouridine (s^2U); 10, 5-methylaminomethyl-2-thiouridine (mnm^5s^2U); 11, inosine (I); 12, 5-methoxycarbonylmethyluridine (mcm^5U); 13, unidentified modification. *E. coli* anticodons, including those from DNA sequences, are from Sprinzl *et al.* (1989); Komine *et al.* (1990). Fungal *mitochondrial* anticodons (*S. cerevisiae*, *N. crassa*, *Aspergillus nidulans*, and *Schizosaccharomyces pombe*), including those from DNA sequences, are from Canaday *et al.* (1980); Heckman *et al.* (1980); Köchel *et al.* (1981); Sibler *et al.* (1986); Dirheimer and Martin (1990).

[c] *Schizosaccharomyces pombe* mitochondria have one species of $tRNA^{Trp}$ with anticodon CCA (not UCA) that translates codon UGA at a low efficiency *in vitro* (Dirheimer and Martin, 1990).

[d] ACG was reported only from *S. cerevisiae* mitochondria (Sibler *et al.*, 1986). UCG is present in mitochondria of *Schizosaccharomyces pombe* (Dirheimer and Martin, 1990) and all animals.

[e] From DNA sequence (Ikeda *et al.*, 1990*b*).

[f] Host nuclear origin (Martin *et al.*, 1979).

[g] ACC (modification unknown) only from *A. nidulans* mitochondria (Köchel *et al.*, 1981).

Table 3.2 Known anticodons in archaebacteria (metabacteria)

Amino acid (codon)	Anticodon in H	Anticodon in M	Amino acid (codon)	Anticodon in H	Anticodon in M	Amino acid (codon)	Anticodon in H	Anticodon in M	Amino acid (codon)	Anticodon in H	Anticodon in M
Phe (UUU)	GAA	GAA	Ser (UCU)	GGA		Tyr (UAU)	GUA	GUA	Cys (UGU)	GCA	
Phe (UUC)			Ser (UCC)			Tyr (UAC)			Cys (UGC)		
Leu (UUA)	UAA CAA		Ser (UCA)			Stop (UAA)			Stop (UGA)		
Leu (UUG)			Ser (UCG)	CGA		Stop (UAG)			Trp (UGG)	CCA	
Leu (CUU)	GAG		Pro (CCU)	GGG		His (CAU)	GUG	GUG	Arg (CGU)	GCG	
Leu (CUC)			Pro (CCC)			His (CAC)			Arg (CGC)		
Leu (CUA)	UAG NAG	UAG	Pro (CCA)	UGG CGG	UGG	Gln (CAA)		UUG	Arg (CGA)	UCG CCG	
Leu (CUG)			Pro (CCG)			Gln (CAG)	CUG		Arg (CGG)		
Ile (AUU)	GAU		Thr (ACU)	GGU	GGU	Asn (AAU)	GUU	GUU	Ser (AGU)	GCU	
Ile (AUC)			Thr (ACC)			Asn (AAC)			Ser (AGC)		
Ile (AUA)	NAU	CAU	Thr (ACA)		UGU	Lys (AAA)	UUU CUU	UUU	Arg (AGA)		UCU
Met (AUG)	CAU		Thr (ACG)	CGU		Lys (AAG)			Arg (AGG)		
Val (GUU)	GAC		Ala (GCU)	GGC		Asp (GAU)	GUC	GUC	Gly (GGU)	GCC	GCC
Val (GUC)			Ala (GCC)			Asp (GAC)			Gly (GGC)		
Val (GUA)		UAC	Ala (GCA)	UGC CGC	UGC	Glu (GAA)	UUC CUC	UUC	Gly (GGA)	UCC CCC	
Val (GUG)	CAC		Ala (GCG)			Glu (GAG)			Gly (GGG)		

Modifications of the anticodon first nucleoside are not indicated. Abbreviations: H, *Halobacterium*; M, methanogens. All methanogen anticodons are from *Methanococcus* sp. except GCC (Gly), which was reported only from *Methanobacterium* sp. (data from Sprinzl *et al.* (1989), reprinted from Osawa *et al.* (1992)).

Table 3.3 Anticodons in chloroplasts

Amino acid (codon)	Anticodon	Amino acid (codon)	Anticodon	Amino acid (codon)	Anticodon	Amino acid (codon)	Anticodon
Phe (UUU)	GAA	Ser (UCU)	GGA	Tyr (UAU)	GUA	Cys (UGU)	GCA
Phe (UUC)		Ser (UCC)		Tyr (UAC)		Cys (UGC)	
Leu (UUA)	UAA CAA	Ser (UCA)	UGA	Stop (UAA)		Stop (UGA)	
Leu (UUG)		Ser (UCG)		Stop (UAG)		Trp (UGG)	CCA
Leu (CUU)	UAG	Pro (CCU)	UGG	His (CAU)	GUG	Arg (CGU)	ACG
Leu (CUC)		Pro (CCC)		His (CAC)		Arg (CGC)	
Leu (CUA)		Pro (CCA)		Gln (CAA)	UUG	Arg (CGA)	
Leu (CUG)		Pro (CCG)		Gln (CAG)		Arg (CGG)	CCG
Ile (AUU)	GAU	Thr (ACU)	GGU	Asn (AAU)	GUU	Ser (AGU)	GCU
Ile (AUC)		Thr (ACC)		Asn (AAC)		Ser (AGC)	
Ile (AUA)	CAU	Thr (ACA)	UGU	Lys (AAA)	UUU	Arg (AGA)	UCU
Met (AUG)	CAU	Thr (ACG)		Lys (AAG)		Arg (AGG)	
Val (GUU)	GAC	Ala (GCU)	UGC	Asp (GAU)	GUC	Gly (GGU)	GCC
Val (GUC)		Ala (GCC)		Asp (GAC)		Gly (GGC)	
Val (GUA)	UAC	Ala (GCA)		Glu (GAA)	UUC	Gly (GGA)	UCC
Val (GUG)		Ala (GCG)		Glu (GAG)		Gly (GGG)	

Modifications are not indicated. Anticodon CCG for Arg was reported only for liverwort chloroplasts (Ohyama *et al.*, 1986; Ozeki *et al.*, 1987; Umesono and Ozeki, 1987; Ohyama *et al.*, 1988); it is not present in tobacco (Shinozaki *et al.*, 1986; Wakasugi *et al.*, 1986) and rice (Hiratsuka *et al.*, 1989) chloroplast genomes. Anticodon GGG for Pro is present in a pseudogene in fern chloroplast genome, so the only anticodon for Pro is presumably UGG (Ozeki *et al.*, 1987).

Table 3.4 Anticodons in vertebrate mitochondria

Amino acid (codon)	Anticodon	Amino acid (codon)	Anticodon	Amino acid (codon)	Anticodon	Amino acid (codon)	Anticodon
Phe (UUU)	GAA	Ser (UCU)	UGA	Tyr (UAU)	GUA	Cys (UGU)	GCA
Phe (UUC)		Ser (UCC)		Tyr (UAC)		Cys (UGC)	
Leu (UUA)	UAA	Ser (UCA)		Stop (UAA)		Trp (UGA)	UCA
Leu (UUG)		Ser (UCG)		Stop (UAG)		Trp (UGG)	
Leu (CUU)	UAG	Pro (CCU)	UGG	His (CAU)	GUG	Arg (CGU)	UCG
Leu (CUC)		Pro (CCC)		His (CAC)		Arg (CGC)	
Leu (CUA)		Pro (CCA)		Gln (CAA)	UUG	Arg (CGA)	
Leu (CUG)		Pro (CCG)		Gln (CAG)		Arg (CGG)	
Ile (AUU)	GAU	Thr (ACU)	UGU	Asn (AAU)	GUU	Ser (AGU)	GCU
Ile (AUC)		Thr (ACC)		Asn (AAC)		Ser (AGC)	
Met (AUA)	CAU	Thr (ACA)		Lys (AAA)	UUU	Stop (AGA)	
Met (AUG)		Thr (ACG)		Lys (AAG)		Stop (AGG)	
Val (GUU)	UAC	Ala (GCU)	UGC	Asp (GAU)	GUC	Gly (GGU)	UCC
Val (GUC)		Ala (GCC)		Asp (GAC)		Gly (GGC)	
Val (GUA)		Ala (GCA)		Glu (GAA)	UUC	Gly (GGA)	
Val (GUG)		Ala (GCG)		Glu (GAG)		Gly (GGG)	

From Barrell *et al.* (1979, 1980) and Osawa *et al.* (1992), with modifications.

Amino acid (codon)	Anticodon	Amino acid (codon)	Anticodon	Amino acid (codon)	Anticodon	Amino acid (codon)	Anticodon
Phe (UUU)	GAA	Ser (UCU)	IGA	Tyr (UAU)	GUA	Cys (UGU)	GCA
Phe (UUC)		Ser (UCC)		Tyr (UAC)		Cys (UGC)	
Leu (UUA)	UAA*	Ser (UCA)	UGA	Stop (UAA)		Stop (UGA)	
Leu (UUG)	CAA	Ser (UCG)	CGA	Stop (UAG)		Trp (UGG)	CCA
Leu (CUU)	IAG	Pro (CCU)	IGG	His (CAU)	GUG	Arg (CGU)	ICG
Leu (CUC)		Pro (CCC)		His (CAC)		Arg (CGC)	
Leu (CUA)	UAG	Pro (CCA)	UGG	Gln (CAA)	UUG	Arg (CGA)	UCG
Leu (CUG)	CAG	Pro (CCG)	CGG	Gln (CAG)	CUG	Arg (CGG)	CCG
Ile (AUU)	IAU	Thr (ACU)	IGU	Asn (AAU)	GUU	Ser (AGU)	GCU
Ile (AUC)		Thr (ACC)		Asn (AAC)		Ser (AGC)	
Ile (AUA)	UAU	Thr (ACA)	UGU	Lys (AAA)	UUU	Arg (AGA)	UCU
Met (AUG)	CAU	Thr (ACG)	CGU*	Lys (AAG)	CUU	Arg (AGG)	CCU
Val (GUU)	IAC	Ala (GCU)	IGC	Asp (GAU)	GUC	Gly (GGU)	GCC
Val (GUC)		Ala (GCC)		Asp (GAC)		Gly (GGC)	
Val (GUA)	UAC	Ala (GCA)	UGC	Glu (GAA)	UUC	Gly (GGA)	UCC
Val (GUG)	CAC	Ala (GCG)	CGC	Glu (GAG)	CUC	Gly (GGG)	CCC

Modifications of the anticodon first nucleoside are not indicated except for I (inosine). UUA and CUA are anticodons for Gln in ciliated protozoa and presumably in *Acetabularia* spp. Anticodons in this table are those found in various eukaryotes; this does not mean that all of them exist in a single species. * Anticodons presumed to exist (data from Sprinzl *et al.* (1989) and Osawa and Jukes (1988), reprinted from Osawa *et al.* (1992)).

4 Codon usage

It should be safe to assume that two synonymous codons in a 2-codon set, or four in a family box, are used evenly. However, there are almost no such examples. Synonymous codon usage is not 'symmetrical'. It is more or less, and sometimes extremely, uneven in any gene in a single species. Several factors affect codon usage, one of which is the genomic G + C content of the organism, another is the amount and composition of isoacceptor tRNAs that translate the codons.

4.1 Genomic G + C contents and codon usage

4.1.1 *Directional mutation pressure*

There are believed to be some 30 000 000 species of organisms on Earth. A large proportion of these belongs to the bacterial kingdom, which includes many species differing widely in their genomic G + C content (ranging from 25 per cent (*Mycoplasma capricolum*) to 75 per cent (*Micrococcus luteus*). A high G + C content is often claimed to be advantageous, e.g. for thermophilicity or halophilicity (Kagawa *et al.*, 1984; Bernardi *et al.*, 1985; Bernardi and Bernardi, 1986; Bernardi, 1989). However, the genomic G + C content is characteristic to each organism and is closely related to phylogeny (Osawa *et al.*, 1990*b*,*c*, 1992). This in turn suggests that the genomic G + C content has changed during evolution. It is therefore more likely that directional mutation pressure results from mutations and that the magnitude of this pressure varies among phylogenetic lines (Fig. 4.1). According to Sueoka (1962), the G + C content of genomic DNA is determined by the effective base conversion rate u (G : C to A : T) and v (A : T to G : C); the G + C content at equilibrium (p) is $v(u + v)$. It follows that directional mutation pressure towards A : T predominating over G : C (AT-pressure) or towards G : C over A : T (GC-pressure) has been exerted on DNA. The most probable cause of the directional mutations is copy errors during DNA replication. The copy errors predominantly from A : T to G : C or from G : C to A : T result in a different genomic G + C content between phylogenetic lines, even if the error frequency is low. In *E. coli*, the mutator gene (*mut*) products are mostly components participating in DNA replication or repair. A mutation of the *mut*T gene specifically induces transversions from A : T to C : G pairs at a high rate (Cox and Yanofsky, 1967); mutation of the other *mut* genes, such as *mut*Y, induces transversions from C : G to A : T (Nghiem *et al.*, 1988). A small amount of a mutagen (8-oxoGTP) in the nucleotide pool induces transversion from A : T to C : G. The *mut*T protein is a nucleoside triphosphatase that decomposes the

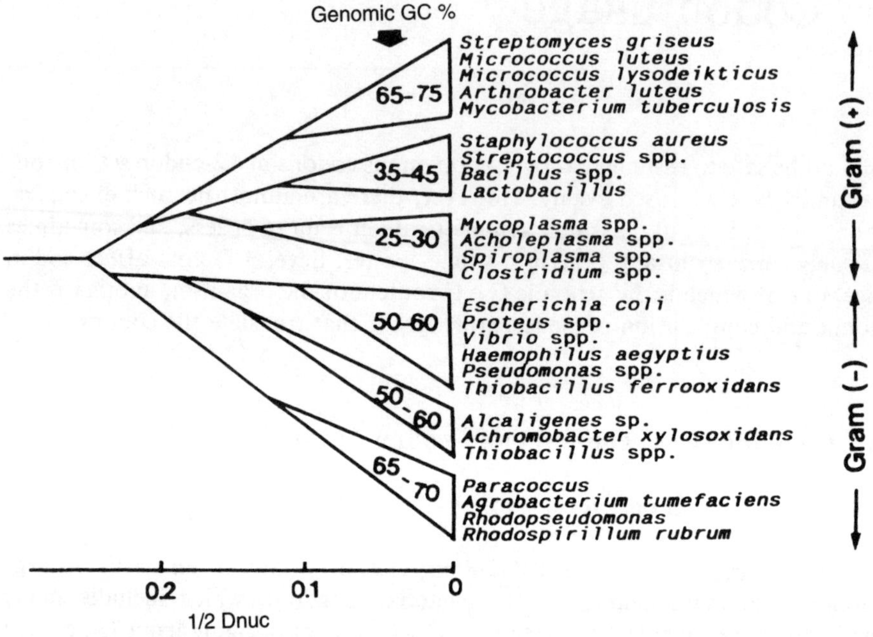

Fig. 4.1. Genomic G + C content is related to phylogeny in eubacteria. Phylogenetic tree of eubacteria was constructed by 5S rRNA sequences. Dnuc, evolutionary distance. The G + C content (%) is given for representative bacteria (from Osawa *et al.*, 1992).

mutagen, so the transversional errors would be suppressed (Akiyama *et al.*, 1989; Maki and Sekiguchi, 1992; see Grollman and Moriya, 1993 for a review). Sueoka (1993) suggested that a sudden change of mutational bias of neutral nucleotide position by the mutation of a mutator gene would cause the G + C content of neutral nucleotides to change.

Thus, it is highly probable that mutators play an important role in the genesis of directional mutation pressure.

4.1.2 *Bacteria*

Figure 4.2 shows the correlation between total genomic G + C contents of various eubacterial species and the spacers, tRNA and rRNA (stable RNA) genes, and protein genes (Muto and Osawa, 1987). There is a weak but positive correlation between the G + C content of the stable RNA genes and the genomic G + C. The G + C content of spacers and protein genes reveals a strong correlation to genomic G + C. Thus, for a given species, the G + C content of these regions are all biased in the same direction as the G + C content of the total genome.

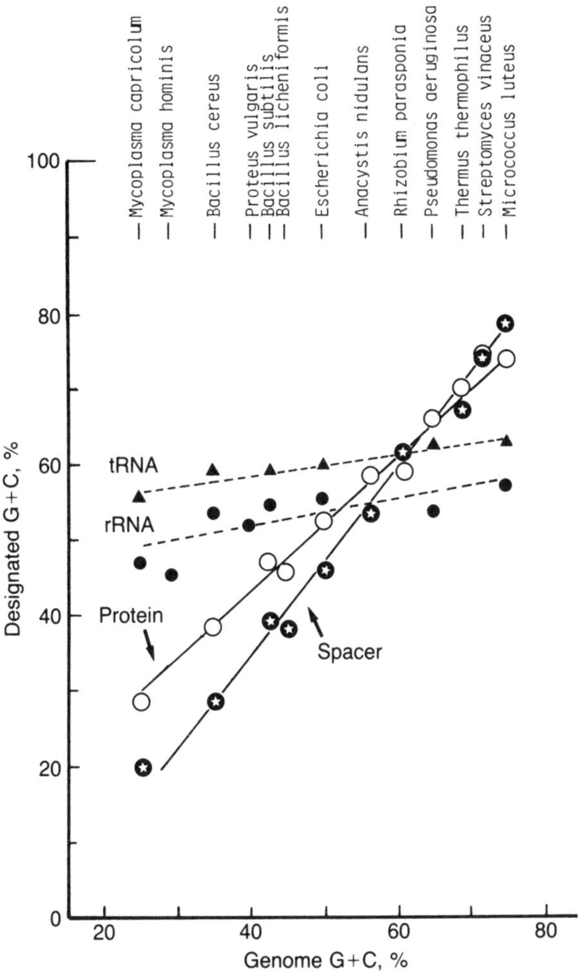

Fig. 4.2. Correlation of G + C content between total genomic DNA and designated parts of the genome (from Muto and Osawa, 1987).

Such mutation pressure seems to be exerted on the entire genome. The different levels of G + C content in the different components of a given organism may be the consequence of selective constraints that have been exerted to eliminate deleterious mutants. As most parts of spacers are functionally the least important in the genome, most mutations in these regions are selectively neutral, and therefore have the highest evolutionary rate. The stable RNA genes are less variable because most of their sequences are important for functions. Protein genes are more variable than the stable RNA genes because of the many synonymous or conservative codon changes. Figure 4.3 shows the correlation between the G + C content of total genomic DNA with that of spacers, for the first,

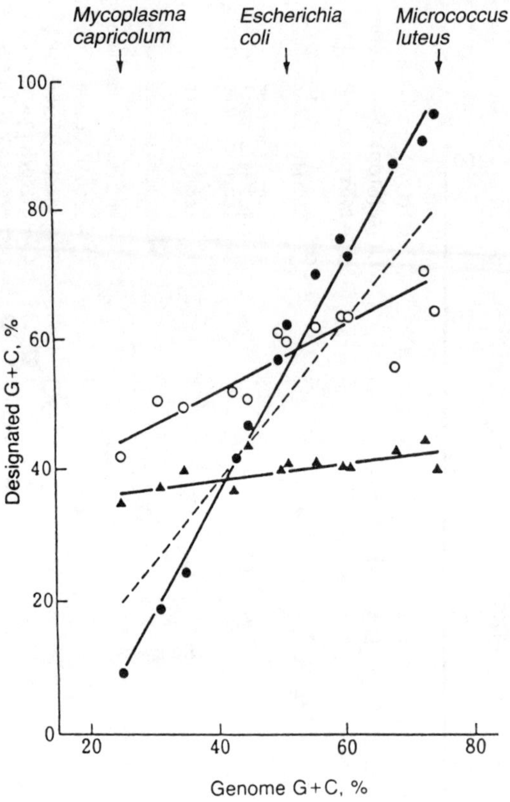

Fig. 4.3. Correlation of the G + C content between total genomic DNA and the first (○), second (▲), and third (●) codon positions of various bacterial species. The positions of three bacterial species are indicated by arrow. The dotted line indicates the percentage spacer G + C level (from Muto and Osawa (1987), with modifications).

second, and third codon positions (Muto and Osawa, 1987). The G + C content of spacers and the third codon position reveals a strong positive correlation with the genomic G + C, because spacers are largely constraint-free, and transition in the third codon position in 2-codon sets [(except for codons AUR (Ile and Met) and UGR (Trp and stop)], and both transitions and transversions in family boxes, are all silent. The first position is next variable, because the synonymous codons for UUR and CUR (both Leu) and for CGR and AGR (both Arg) differ in the first position. Many conservative (neutral or near-neutral) substitutions occur in the first position. Changes in the second position always result in amino acid replacements, so that functional constraint is the strongest in this position. This order coincides with that of the differences in variability or the relative occurrence of neutral substitutions at the three codon positions, implying that most of the directional substitutions are neutral or near-neutral.

Table 4.1 Codon usage in *Micrococcus luteus*, *Escherichia coli*, and *Mycoplasma capricolum*

Codon	L	E	M	Codon	L	E	M	Codon	L	E	M	Codon	L	E	M
UUU (Phe)	<1	19	39	UCU (Ser)	1	11	16	UAU (Tyr)	<1	14	25	UGU (Cys)	<1	4	6
UUC (Phe)	31	17	4	UCC (Ser)	30	10	<1	UAC (Tyr)	25	14	4	UGC (Cys)	4	6	1
UUA (Leu)	0	10	64	UCA (Ser)	<1	6	25	UAA (Stop)	—	—	—	UGA (Stop)	—	—	6*
UUG (Leu)	1	11	3	UCG (Ser)	13	7	1	UAG (Stop)	—	—	—	UGG (Trp)	4	11	1
CUU (Leu)	<1	9	5	CCU (Pro)	17	3	<1	CAU (His)	1	11	11	CGU (Arg)	10	28	8
CUC (Leu)	35	9	0	CCC (Pro)	2	6	10	CAC (His)	18	11	3	CGC (Arg)	50	21	1
CUA (Leu)	0	3	10	CCA (Pro)	<1	7	18	CAA (Gln)	0	13	38	CGA (Arg)	1	3	<1
CUG (Leu)	50	57	<1	CCG (Pro)	27	25	<1	CAG (Gln)	39	31	1	CGG (Arg)	15	4	0
AUU (Ile)	<1	26	70	ACU (Thr)	1	11	30	AAU (Asn)	1	15	65	AGU (Ser)	<1	6	17
AUC (Ile)	53	30	8	ACC (Thr)	43	24	1	AAC (Asn)	28	25	11	AGC (Ser)	4	15	3
AUA (Ile)	0	3	16	ACA (Thr)	<1	6	22	AAA (Lys)	<1	38	107	AGA (Arg)	0	1	28
AUG (Met)	22	26	22	ACG (Thr)	23	11	<1	AAG (Lys)	51	12	10	AGG (Arg)	2	1	<1
GUU (Val)	<1	23	42	GCU (Ala)	2	19	33	GAU (Asp)	3	31	43	GGU (Gly)	9	31	26
GUC (Val)	44	14	1	GCC (Ala)	54	23	<1	GAC (Asp)	51	23	4	GGC (Gly)	68	31	1
GUA (Val)	0	13	23	GCA (Ala)	3	21	21	GAA (Glu)	<1	46	55	GGA (Gly)	2	5	29
GUG (Val)	52	25	2	GCG (Ala)	28	34	1	GAG (Glu)	73	19	4	GGG (Gly)	9	9	2

The numbers show the codon usage in the protein genes expressed in frequency per thousand codons.
< 1 represents a codon frequency of less than 0.4. L: *Micrococcus luteus*; E: *Escherichia coli*; M: *Mycoplasma capticolum*.
*UGA is Trp codon in *M. capricolum*.
From Osawa *et al.* (1990*a*).

Comparisons of the codon usage of *Micrococcus luteus* (genomic G + C content = 75 per cent), *E. coli* (50 per cent), and *Mycoplasma capricolum* (25 per cent) also confirm our view (Table 4.1). The G + C content of the third codon position is more than 95 per cent in *Micrococcus luteus*, 53 per cent in *E. coli*, and 10 per cent in *Mycoplasma capricolum*. The same bias is seen at the first positions. Among six synonymous codons of Leu, *Micrococcus luteus* almost exclusively uses CUC and CUG codons, whereas *Mycoplasma capricolum* prefers UUA codons. A similar situation exists for the Arg synonymous codons CGC and CGG (*Micrococcus luteus*) and AGA (*Mycoplasma capricolum*).

The above view may also be applied to initiation and stop codons. In fact, GUG is used as an initiation codon in 10 of 18 protein genes examined in *Micrococcus luteus* (genomic G + C = 75 per cent; Ohama *et al.*, 1989), whereas AUG greatly predominates over GUG in *E. coli* (G + C = 50 per cent) and is used exclusively in *Mycoplasma capricolum* (G + C = 25 per cent) (Ohkubo *et al.*, 1987). Most termination codons in *Mycoplasma capricolum* and *E. coli* are UAA,

Table 4.2 Initiation and termination codons in *Micrococcus luteus*, *Mycoplasma capricolum*, and *Escherichia coli*

Protein	*M. luteus*	*E. coli*	*M. capricolum*
L14	GUG (UGA)	AUG (UAA)	AUG (UAG)
L24	AUG (UGA)	AUG (UAA)	AUG (UAA)
L5	AUG/GUG[a] (UAA)	AUG (UAA)	AUG (UAG)
S14	ND[b] (ND)	AUG (UAG)	AUG (UAG)
S8	AUG (UGA)	AUG (UAA)	AUG (UAA)
L6	AUG (UGA)	AUG (UAA)	AUG (UAG)
L18	GUG (UGA)	AUG (UAA)	AUG (UAA)
S5	GUG (UGA)	AUG (UAA)	AUG (UAA)
L30	GUG (UGA)	AUG (UAA)	ND (ND)
L15	AUG (UGA)	AUG (UAA)	AUG (UAA)
secY	GUG (UGA)	AUG (UAA)	AUG (UAA)
X	ND (ND)	AUG (UGA)	ND (ND)
adk	AUG (UGA)	AUG (UAA)	AUG (UAA)
S12	GUG (UAA)	AUG (UAA)	ND (ND)
S7	AUG (UGA)	AUG (UGA)	ND (ND)
EF-G	GUG (UGA)	AUG (UAA)	ND (ND)
EF-Tu	GUG (UGA)	GUG (UAA)	ND (ND)
uvrB	AUG (UGA)	AUG (UAA)	ND (ND)
uvrA	GUG (UAG)	AUG (UAA)	ND (ND)

Initiation codons and termination codons (in parentheses) of the known *M. luteus* genes and those of corresponding *E. coli* and *M. capricolum* genes.
[a] Initiation codon of the *M. luteus* L5 gene is either AUG or GUG.
[b] Not determined.
The initiation codon of an *M. luteus* ORF after *adk* is also GUG.
From Ohama *et al.* (1989).

whereas *Micrococcus luteus* uses 15 UGA, 2 UAA, and 1 UAG (Ohama *et al.*, 1989) (Table 4.2). Brown *et al.* (1990*a,b*) compared the usage of three stop codons in prokaryotes with the genomic G + C content and found that, the higher the A + T content, the more usage of UAA. These observations are in accordance with the positive relationship between the G + C content of genomic DNA and that of codons (Fig. 4.3).

The relationship between the G + C content of the genomes and that of the three codon positions was also studied by Bernardi and Bernardi (1986) for prokaryotic, viral, and vertebrate genes. The results are practically the same as those mentioned above, but Bernardi and Bernardi gave an entirely different explanation, i.e. that the bias is the result of environmental pressures, especially temperature. However, as discussed by Filipski (1991):

> Among both thermophilic and mesophilic organisms there are some with G + C-rich genomes and others with AT-rich genomes. This would suggest that a high genomic GC-content is irrelevant to the adaptation of species to environments of high temperature . . . No evidence has been found for (or against) the evolutionary enrichment in GC of genomes of those species which live in an environment of elevated temperature.

4.1.3 *Eukaryotes*

Systematic analyses of the relation between the genomic G + C content and codon usage in various organisms other than eubacteria, which are described briefly below, are rather fragmentary, and yet extensive analyses by Sueoka (1988) using the codon usage database of both unicellular and multicellular organisms gave the same conclusion for eubacteria. Chloroplasts from tobacco (Shinozaki *et al.*, 1986), liverwort (Ohyama *et al.*, 1986), and rice; mitochondria from yeasts, liverwort (Ohyama *et al.*, 1991), and maize; and some lower eukaryotes such as *Paramecium* spp. (Preer *et al.*, 1985; Prat *et al.*, 1986) and slime mould (Warrick and Spudich, 1988) have genomes with a low G + C content and reveal codon usage patterns similar to those of *Mycoplasma capricolum*. The codon usage pattern of the *Acanthamoeba castellanii* actin gene (Nellen and Gallwitz, 1982) closely resembles that of *Micrococcus luteus*. In the high G + C alga *Chlamydomonas reinhardtii*, there is a strong G + C influence on silent sites of codons in 14 genes examined (Brown *et al.*, 1990*b*).

The chromosomes of higher vertebrates reveal mosaic structures consisting of G + C-rich and G + C-poor DNA segments, which correspond to R bands and G bands, respectively (Bernardi *et al.*, 1985; Aota and Ikemura, 1986; Ikemura and Aota, 1988; Bernardi, 1989; Holmquist, 1989). In accordance with this, the silent positions of codons are high in G and C in the G + C-rich segments, and low in G and C in the G + C-poor segments. DNA replication in eukaryotes takes place in two phases; the R-band DNAs replicate early in the cell cycle and the G-band DNAs replicate late in the cell cycle. GC-pressure might be higher in the early phase, whereas AT-pressure might predominate in the late phase because of the possible use of two different replication systems. For further discussions of directional mutation pressure and the origin of chromosomal compartments, see the review by Filipski (1991).

4.1.4 *Mitochondria*

Generally, one strand of DNA has, on average, about the same base composition as the other strand, which suggests that directional mutations and their fixations occur evenly on both the strands. This is seen in the genomes of many organisms and in certain mitochondria, such as those of green plants (Ohyama *et al.*, 1991), yeasts (Zamaroczy and Bernardi, 1986), *Trypanosoma brucei* (Hensgens *et al.*, 1984), *Drosophila yakuba* (Clary and Wolstenholme, 1985), and *Mytilus edulis* (Hoffmann *et al.*, 1992). So T (or C) contents are nearly equal to A (or G) contents in the silent sites of codons, regardless of which strand codes for the gene. The order of choice in the both L and H strands in *Drosophila* is $T = A \geqslant C = G$, while in *Mytilus* it is $T \fallingdotseq A > C \fallingdotseq G$. In vertebrate mitochondria the G + C content is uneven between the H and L strands. The G + C (mostly C) content of codon silent sites found mostly on the L strand increases in species from 28 per cent (*Xenopus laevis*) to 51 per cent (human) (Jukes and Bhushan, 1986). This may be due to different degrees of selective mutations and/or to fixations of one of the two DNA strands. During replication the H strand seems to be exposed as a single strand for much longer than the L strand. As there is therefore less chance for repair, observable mutations accumulate more in the H strand than in the L strand. In fact, there is marked gradient along vertebrate mitochondrial genomes in contents of T- and C-ending codons in the L strand (Niko *et al.*, unpublished data). The selective mutations on the H strand may be A to G, leading to accumulation of C in the major sense (mRNA-like) L strand by replacement of T by C (Anderson *et al.*, 1982). This accumulation of C occurs in the silent sites of codons and also in the D loop, tRNA-like, and rRNA-like sequences in the L strand. The order in L strand is $C \fallingdotseq A \geqslant T > G$. Thus, GC-pressure, which in this case is somewhat different from the usual symmetrical GC-pressure, acts on the whole genome. Interestingly, the silent sites of the NADH dehydrogenase subunit 6 sequence in the H strand are rich in G because of loss of A, in contrast to the richness of C in other mRNA-like sequences on the L strand (Anderson *et al.*, 1982).

The pattern of such asymmetrical directional mutations seems to be specific to mitochondria of certain lineages. In echinoderm mitochondria (Himeno *et al.*, 1987; Jacobs *et al.*, 1988; Cantatore *et al.*, 1989; Asakawa *et al.*, 1991), the order of choice in the L strand is $A \gtrsim T = C > G$, whereas in the mitochondria of *Fasciola hepatica* (Garey and Wolstenholme, 1989) and *Dugesia japonica* (planaria) (Bessho *et al.*, 1991) the order in the mRNA-like strand is $T > G \geqslant A \fallingdotseq C$, which is nearly the reverse of the order in the vertebrate mitochondrial L strand. The mRNA-like sense strand of platyhelminths could be an H strand. These asymmetrical directional mutation pressures could be a result of the difference in the degree of repair and proofreading between the L and H strands.

4.2 Codon selection by tRNA

The G + C content of the silent site of codons is determined primarily by directional mutation pressure and is modulated to varying degrees by selection pressure, mainly by tRNAs. The fact that the slope of G + C content of the third codon position is steeper than that of spacers (see Fig. 4.3) may be owing to the contribution of tRNAs as described below.

Ikemura (1981*a*,*b*; 1982) found a close correlation between the intracellular abundance of tRNAs and the occurrence of various codons in *E. coli* and *Saccharomyces cerevisiae*. He stated that:

1. When synonymous codons for an amino acid are translated by more than two tRNAs with different anticodons (isoacceptors), the codons recognized by the abundant tRNA species are used more frequently than those read by the less abundant tRNA(s). For example, the abundance of the Leu codon CUG among six synonymous Leu codons in *E. coli* is accompanied by the largest intracellular amount of the corresponding $tRNA^{Leu}_{CAG}$ among the isoacceptors. The 'optimal' codons differ between organisms because populations of isoacceptors differ.
2. When one species of tRNA anticodon reads more than one codon by wobbling, the higher the pairing affinity of the codon to the anticodon, the higher the usage. Thus, in NNY-type 2-codon sets, the NNC codon is used more frequently than the NNU, because anticodon GNN forms a more stable base-pair with codon NNC than with codon NNU.

Ikemura (1981*b*) concluded that bias of synonymous codon usage is determined by the relative amounts of isoacceptor tRNAs and/or by the nature of the anticodons, acting as a selection pressure. However, this does not include the primary effect of directional mutation pressure.

Inefficient translation of codons by rare anticodons is subject to negative selection. Therefore, the codons used in the highly expressed genes are more efficiently selected against. In effect, 'optimal' codons are positively selected by tRNAs, as in the case of CUG Leu codon in *E. coli*. In weakly expressed genes, the major factor determining synonymous codon usage is mutation pressure, because the effect of tRNAs is weak or absent owing to a relaxation of positive selection by tRNA. Figure 4.4 shows the correlation between the relative usage of NNC and NNU codons in NNY-type 2-codon sets, and the expression levels of more than 200 different genes in *E. coli* (Ohama *et al.*, 1990*a*). As the two codons in the sets are each translated by a single anticodon, GNN, the positive selection by tRNAs is due exclusively to the higher affinity between codon and anticodon. The higher the expression level of the genes, the higher the usage of NNC codons. The bias decreases with decreasing expression level and reaches a 'bottom' value of about 40 per cent in *E. coli* and 34 per cent in *B. subtilis*. The bottom value may be taken as an approximate measure of the level of directional mutation pressure.

The above description does not mean that all the silent positions in weakly expressed genes are free from such positive selection, or that those in highly

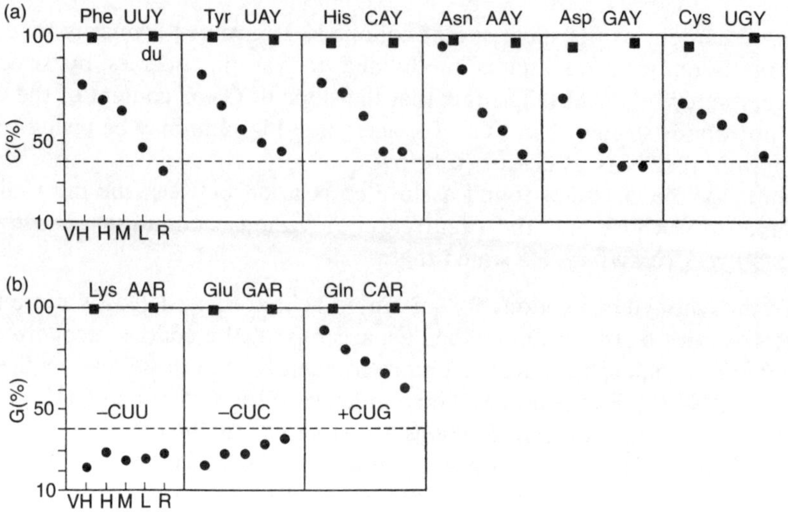

Fig. 4.4. Codon usage in 2-codon sets of various classes of genes in *E. coli* (●) and *Micrococcus luteus* (■). VH, very highly expressed genes; H, highly expressed genes; M, moderately expressed genes; L, poorly expressed genes; R, regulatory genes which are expressed poorly. (a) Usage of NNC codons (%) in NNY-type 2-codon sets; the dashed line represents the average NNC content of the class R genes of all the NNY-type 2-codon sets. (b) Usage of NNG codons in NAR-type 2-codon sets; + or − CUN means the presence or absence of CUN anticodon in translation of the 2-codon sets. In *Micrococcus luteus* du represents *dnaA* and *uvrB* genes (from Osawa *et al.*, 1992).

expressed genes are positively selected by tRNAs. The situation is very different in the NNR-type. For example, in both *E. coli* and *B. subtilis*, contents of NNG codons in NNR-type 2-codon sets (AAR Lys and GAR Glu) are less than 35 per cent when translated by a single anticodon, *UNN, regardless of the expression levels. The NNG content in *E. coli* is much less than that of spacer regions (47 per cent), suggesting the presence of negative selection by *UNN anticodons to use fewer NNG codons than NNA throughout the genes. This would occur because anticodon *UNN pairs mainly with codon NNA and very poorly with NNG, so that, even in weakly expressed genes, NNG cannot be used extensively. On the other hand, the presence of the CNN anticodon (e.g. CUG for Gln), which exclusively translates codon NNG by forming a strong C1 (anticodon) : G3 (codon) pair, greatly enhances the usage of NNG, with a gradient from the highly to weakly expressed genes. The NNG content in a weakly expressed gene is 60 to 70 per cent, which is higher than the spacer G+C content, suggesting that positive selection by CNN anticodons to use more NNG codons has been exerted even in this class of genes. The development of CNN anticodons would respond to a tendency to use more NNG codons to avoid negative selection. In fact,

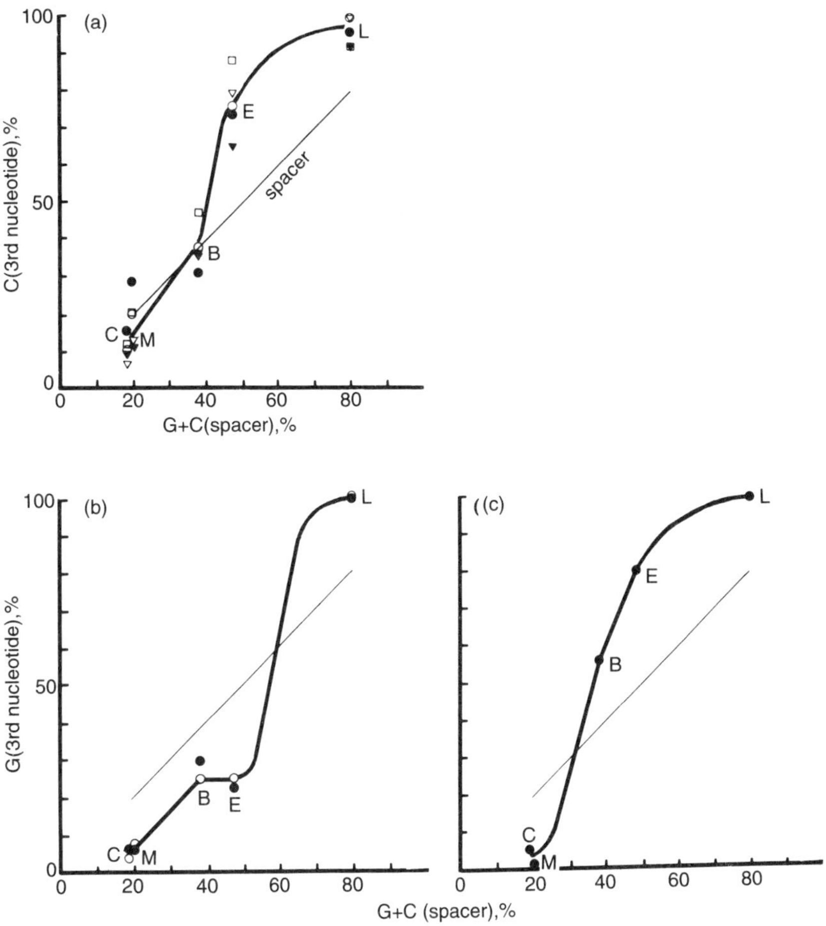

Fig. 4.5. Relation between G + C content of spacer and (a) C or (b and c) G content of the third nucleotide of 2-codon sets. (a) Codons for Tyr (○), His (●), Asn (□), Asp (▼), and Phe (▽). (b) Codons for Lys (○) and Glu (●). (c) Codon for Gln (●). B, *Bacillus subtilis*; C, *Marchantia polymorpha* chloroplast; E, *Escherichia coli*; L, *Micrococcus luteus*; M, *Mycoplasma capricolum*. The straight lines indicate % spacer G + C levels (45° slope) (from Osawa *et al.*, 1988).

bacteria with high G + C contents have CNN anticodons in NNR sets (Osawa and Jukes, 1988; Osawa *et al.*, 1990*b*,*c*).

The phenomena described above become more apparent when synonymous codon usages in 2-codon sets of highly expressed genes are examined for various bacteria with widely different genomic G + C contents (Osawa *et al.*, 1988) (Fig. 4.5). In NNY-type 2-codon sets, usage of NNC codons becomes higher than the C content of spacers with increasing genomic G + C content, suggesting that the higher the genomic G + C content (GC-pressure), the greater the positive selection of NNC codons by GNN anticodons. These facts imply that when the C content of the codon third positions increases (by GC-pressure), translation efficiency will decrease because of the occasional presence of NNU as a rate-limiting factor. This will result in positive selection of NNC codons by GNN anticodons and an increase in the number of NNC codons. On the other hand, in bacteria (including chloroplasts) with low G+C (or high A+T) genomes, NNC usage does not differ from the spacer C content; occasional occurrence of NNC codons would not greatly affect the overall translation efficiency.

Anticodon CNN for codon NNG does not exist in NNR-type 2-codon sets of A+T-rich bacteria. The sole anticodon *UNN for codon NNR does not pair well with NNG codon, so that positive selection to use more NNA than NNG would occur. With increasing GC-pressure, NNG usage is subjected to negative selection by anticodon *UNN, as noted above (AAG Lys and GAG Glu of *Mycoplasma capricolum*, fern chloroplasts, *B. subtilis*, and *E. coli*). Development of CNN anticodon greatly enhances NNG usage over G usage in spacers (AAG Lys, and GAG Glu of *Micrococcus luteus*; CAG Gln of *B. subtilis*, *E. coli* and *M. luteus*, although anticodon CUG Gln has not been reported for *B. subtilis*).

The synonymous codon selection by tRNAs in family boxes is more difficult to analyse, because of complicated quantitative and qualitative interrelations between the two to three isoacceptor tRNAs that usually translate codons in family boxes. Nevertheless, there is a clear tendency for bias by tRNA to be stronger in highly expressed genes than in weakly expressed genes.

In *Micrococcus luteus*, both highly expressed genes (such as ribosomal protein genes) and weakly expressed genes (such as *dna*A and *urv*B) have an average G+C content in the silent codon positions of about 95 per cent (Ohama *et al.*, 1990*a*; Osawa *et al.*, 1990*c*), which is about 20 per cent higher than in the spacers. This may be explained as follows: G+C contents of the silent codon positions of both highly and weakly expressed genes in extremely high G+C bacteria, such as *M. luteus*, are positively biased by anticodon CNN or GNN to varying degrees, so that the G+C content of these positions reaches near-saturation, in contrast to the almost constraint-free spacers (Ohama *et al.*, 1990*a*).

4.3 Stop codon selection by release factors

As described above, the use of stop codons UAA and UGA is correlated with genomic G+C contents, whereas UAG is used only rarely throughout the eubacteria. Stop codons have no corresponding tRNA. UAA and UAG are recognized by release factor 1 (RF-1), and UAA and UGA are recognized by RF-2. These facts, together with a much lower usage of UAG than of UAA or UGA, suggest that RF-2 has a higher affinity than RF-1 for UAA, so that most UAAs (and all UGAs) would be recognized by RF-2 (Brown *et al.*, 1990*a*). Thus, much higher usage of UAA and UGA over UAG could be brought about by selection pressure because of the apparently higher efficiency of RF-2 for translational termination. UAA predominates over UGA in highly expressed *E. coli* genes, and the use of UGA increases with decrease of expression level, reaching almost the same level of usage as UGA and UAA in poorly expressed genes. This suggests that a positive selection pressure by RF-2 to use UAA has been exerted in the *E. coli* lineage. In *Micrococcus luteus*, UGA greatly predominates over UAA (Ohama *et al.*, 1989), suggesting that GC-pressure is the main force for this usage.

Brown *et al.* (1990*a*) suggest that the nucleotide following the stop codon is important for efficient translational termination. U is the most highly represented in the nucleotide position following all three stop codons in *E. coli*; A and C are less frequent. The situation is similar in other bacteria with a moderate or lower genomic G+C content, such as *Salmonella typhimurium*, *B. subtilis*, *Mycoplasma capricolum*, some methanogenic archaebacteria, and bacteriophages. This pattern is accentuated in highly expressed genes, but is not as marked in either weakly expressed genes or in those that terminate in UAG—the codon recognized only by RF-1. This also suggests that RF-2 selects for stop signal choice in these bacteria, in addition to directional mutation pressure. However, UGA is the most abundant stop codon in *Micrococcus luteus* and *Streptomyces* spp. with high genomic G+C contents (UAA is much less abundant; see above), and 70 to 80 per cent of the nucleotides following stop codons are C or G. In *Micrococcus luteus*, C follows 11 of 18 stop codons examined, with two Gs, one A, and four Us (Ohama *et al.*, 1987, 1989). This pattern is completely different from that of other bacteria. This is probably due to the predominance of high GC-pressure over the selection by RF, so that not only the stop codon, but also the adjacent nucleotide, is influenced mainly by directional mutation pressure.

In eukaryotes, a single RF protein (eRF) recognizes three stop codons (Caskey, 1980). Here again, certain stop codons and nucleotides following stop codons are used preferentially, and the pattern is accentuated in highly expressed genes. The pattern varies, depending on the organism, but generally the signals UAA followed by A or G, and UGA followed by A or G, are preferred (Brown *et al.*, 1990*b*). For example, rabbit eRF recognizes the UGAN in the order UGAG = UGAA >> UGAC > UGAU, which is similar to the frequency of their occurrence. This order is also similar to the order of cross-linking between eRF and UGAN (UGAA = UGAG > UGAU > UGAC) (Tate, cited in Farabaugh, 1993).

5 Unassigned or nonsense codons

5.1 Genesis of unassigned codons from amino acid codons

The genomic G + C content of *Micrococcus luteus* is 75 per cent, the highest of all the organisms that have been examined. Figure 4.1 shows that codons ending with G or C comprise 95 to 100 per cent of all codons (Ohama *et al.*, 1990*a*). In the NNY pair, use of the NNC codon does not reach 100 per cent, and a small amount of NNU is utilized. This is because codons NNC and NNU are normally read by a single anticodon, GNN (except for family boxes in mitochondria and *Mycoplasma capricolum*), so that the NNU codon can be regenerated by back-mutation from NNC (or NNG in family boxes) to some extent. On the other hand, NNG usage is 100 per cent in six out of 14 NNR pairs. In high G+C bacteria, NNR codons are read by two anticodons; NNA by anticodon *UNN (or +UNN in family boxes) and NNG mainly by CNN, and inefficiently by *UNN, especially in 2-codon sets. Therefore, complete conversion of an NNA codon to its synonymous NNG codon by GC-pressure, together with positive selection of NNG by anticodon CNN, is likely to occur if the *UNN anticodon is deleted. If this happens, the NNA codon regenerated by back-mutation from NNG is subjected to negative selection. Even if GC-pressure is weakened during evolution, with an increase in mutation rate from NNG to NNA, NNA cannot appear because of the absence of the *UNN anticodon that translates it. In effect, such an NNA codon cannot exist in the coding frames. Such a codon is called an unassigned or nonsense codon.

Does the deletion of certain anticodons actually happen? A close correlation between codon usage and the amount of isoacceptor tRNA originally found by Ikemura (1981*a*,*b*; 1982) suggests that, under extreme GC-pressure, such a deletion is indeed possible.

Figure 5.1(a) shows the relation between synonymous codon usage and the amount of the corresponding tRNAs (anticodons) in *Micrococcus luteus* (Kano *et al.*, 1991); there is a strong correlation. Generally speaking, a large amount of tRNAs with anticodons GNN and CNN translate the abundantly occurring NNC and NNG codons, whereas the amount of anticodon *UNN is very small or not detectable, in accordance with disuse of NNA codons. It is, then, highly probable that some of these non-detectable tRNAs are really absent.

In principle, the situation is the same in *Mycoplasma capricolum*, which has a very high genomic A + T content (75 per cent) (Yamao *et al.*, 1991) (Fig. 5.1(b)). Here, codons NNU and NNA predominate over codons NNC and NNG. The tRNAs translating these NNU and NNA codons exist in large amounts, whereas anticodons CNN for NNG codons are non-existent, except for $tRNA^{Trp}_{CCA}$ and $tRNA^{Lys}_{CUU}$. Even under a strong AT-pressure, use of codons NNC and NNG in 2-codon sets is seen. This is due to the wobble-reading of NNC and NNU codons by anticodon GNN, and to that of NNA and NNG by anticodon *UNN, enabling back-mutation from NNU to NNC, or from NNG to NNA. Family boxes contain only a single species of tRNA with anticodon UNN (unmodified U) except for Thr (see above) and Arg (CGN). As anticodon UNN can translate all four codons in a family box by 4-way wobbling, small amounts of the codons NNC and NNG are always used.

In *Mycoplasma capricolum*, codon CGG—an Arg codon in the universal genetic code—has not been detected among more than 6000 codons examined (Osawa *et al.*, 1992) (see Table 4.1). Also undetected are the tRNA with anticodon CCG for codon CGG, and its gene (Andachi *et al.*, 1989; Muto *et al.*, 1990). It is therefore probable that, under strong AT-pressure, codon CGG was completely converted to its synonymous codon CGU or CGA (read by anticodon ICG), or to AGR (mainly AGA mutated via AGG; read by anticodon *UCU). As a result, both codon CGG and $tRNA^{Arg}_{CCG}$ would have been removed from the genome.

In the mitochondria of the yeast *Torulopsis glabrata*, which has an A + T-rich genome, neither codon CGN (Arg) nor the corresponding tRNA have been found (Clark-Walker *et al.*, 1985). All Arg codons are AGR (mainly AGA). However, the mitochondria of *Saccharomyces cerevisiae* (a close relative of *T. glabrata* that also has an A + T-rich genome), use a small amount of CGN as Arg codons. These codons seem to be translated by a single species of $tRNA^{Arg}_{ACG}$ (unmodified A) (Sibler *et al.*, 1986). The A + U content of this tRNA is extremely high (81 per cent) (Table 5.1), presumably because of an accumulation of A and U in the tRNA along with conversion of codons CGN to AGA under strong AT-pressure (Osawa *et al.*, 1990*a*).

The absence of CGN codons and their responsible tRNA from *Torulopsis* mitochondria would represent a more advanced stage of the directional changes than observed in *Saccharomyces* mitochondria.

The mitochondrial genome of the chlorophyte alga *Prototheca wickerhamii* is thought to be under high AT-pressure. A and T together comprise 94 per cent of third codon positions in family boxes, corresponding to a high overall A + T content for this genome (74.2 per cent). Complete sequence of this mitochondrial DNA shows that all amino acid codons are universal ones, but that codons TAG, TGA, and CGG are not used at all (Wolff *et al.*, 1994). As the gene for $tRNA^{Arg}_{CCG}$, which translates the CGG codon, does not exist in *P. wickerhamii* mitochondria, CGG is certainly an unassigned codon, presumably produced under high AT-pressure, as in the case of *Mycoplasma capricolum*. UGA is also likely to be an unassigned codon. Only one species of $tRNA^{Trp}$ with anticodon CCA for codon UGG exists in the genome, and indeed, all Trp codons in this mitochondrion are UGG. To decide whether UGA is an unassigned codon, further studies are needed to correlate the absence of UGA with lack of RF-2 activity. If UGA is an unassigned codon, UAG, although it does not exist in the genome, would not

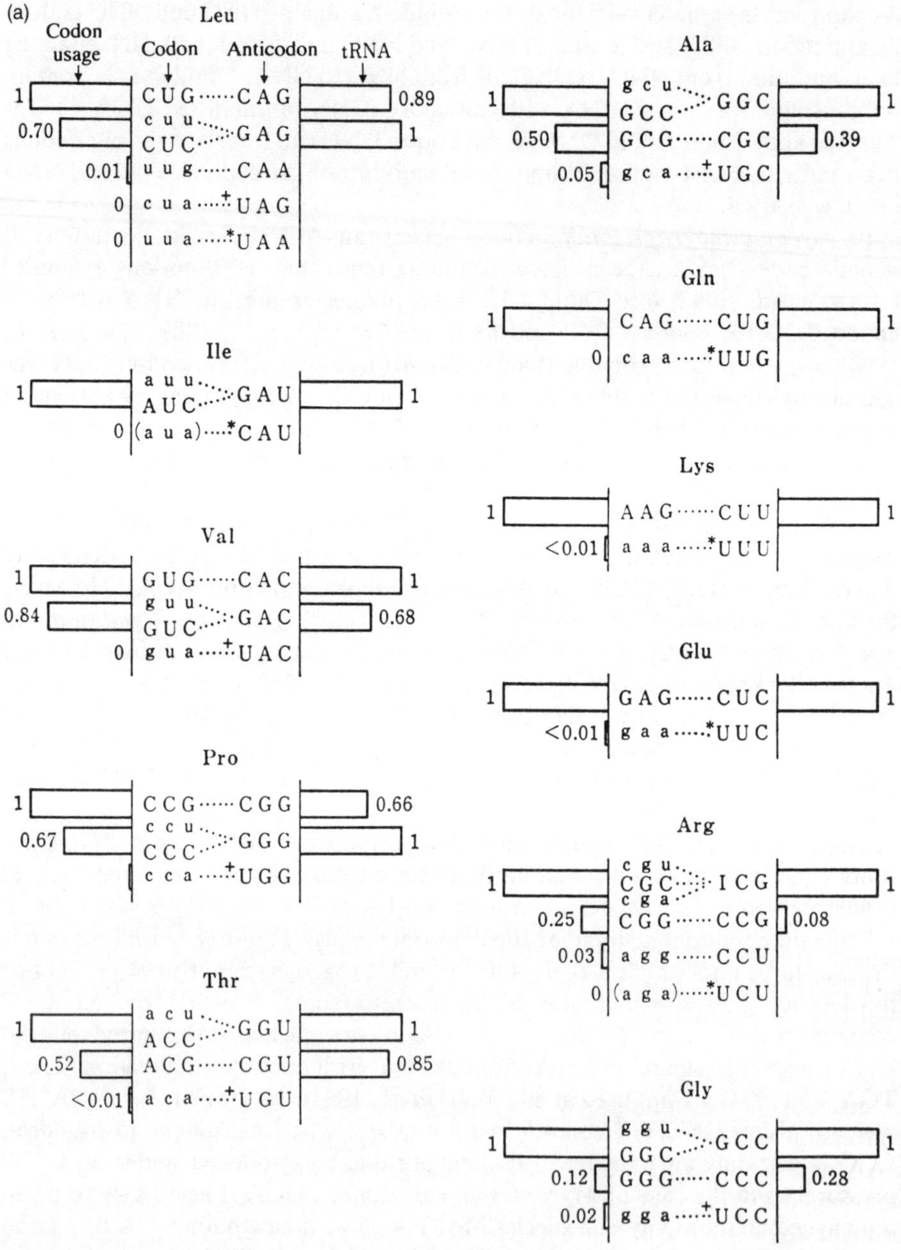
(a)
Leu
Codon usage
Codon
Anticodon
tRNA
1 CUG····· CAG 0.89
0.70 cuu CUC GAG 1
0.01 uug····· CAA
0 cua····+UAG
0 uua····*UAA
Ile
1 auu AUC GAU 1
0 (aua)····*CAU
Val
1 GUG····· CAC 1
0.84 guu GUC GAC 0.68
0 gua····+UAC
Pro
1 CCG····· CGG 0.66
0.67 ccu CCC GGG 1
cca····+UGG
Thr
1 acu ACC GGU 1
0.52 ACG····· CGU 0.85
<0.01 aca····+UGU
Ala
1 gcu GCC GGC 1
0.50 GCG····· CGC 0.39
0.05 gca····+UGC
Gln
1 CAG····· CUG 1
0 caa····*UUG
Lys
1 AAG····· CUU 1
<0.01 aaa····*UUU
Glu
1 GAG····· CUC 1
<0.01 gaa····*UUC
Arg
1 cgu CGC cga ICG 1
0.25 CGG····· CCG 0.08
0.03 agg····· CCU
0 (aga)····*UCU
Gly
1 ggu GGC GCC 1
0.12 GGG····· CCC 0.28
0.02 gga····+UCC

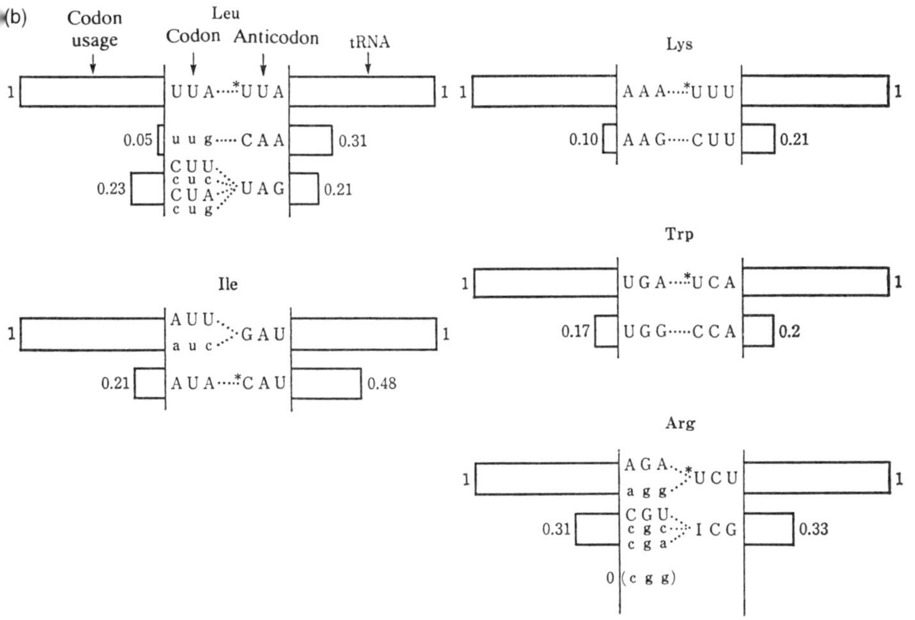

Fig. 5.1. Correlation between codon usage and the relative amount of isoacceptor tRNAs in (a) *Micrococcus luteus* and (b) *Mycoplasma capricolum*. The relative amount of isoacceptor tRNAs was compared with the frequency of choosing synonymous codons that are translatable by the isoacceptor tRNAs. Relative amounts of each codon and anticodon were expressed in the ratio of the most abundant one as 1. All the codons and anticodons are shown, regardless of their usage. Codons in capital letters and lower letters show more than and less than 25 per cent usage in the synonymous codons, respectively. Codon–anticodon pairing is shown by dotted lines, and was deduced from the pairing patterns in other eubacteria. Pairing between codon NNU and anticodon ⁺UNN has been omitted to avoid complexity. See text for modifications of U, ⁺U and *U (redrawn from (a) Kano *et al.* (1991), and (b) Yamao *et al.* (1991)).

be an unassigned codon. Its absence may have been brought about simply by conversion of UAG to UAA by AT-pressure, because RF-1 would recognize both UAA and UAG. Alternatively, it is also possible that UAG (not UGA) is an unassigned codon by the lack of RF-1 activity, while non-use of UGA was simply brought about by AT-pressure having been converted to UAA.

The observation of the apparent deletion of a codon and its corresponding tRNA in *M. luteus*, *M. capricolum*, and *Torulopsis* and *Prototheca* mitochondria may be related to directional mutation pressure, which causes a codon to fall into disuse by converting it to its synonymous codon. The functional constraints on

Table 5.1 Base composition of *Saccharomyces cerevisiae* mitochondrial tRNA genes

tRNA	% G+C
Arg (ACG)	18
Thr (UAG)	21
Met (CAU)	26
Ser (GCU)	28
Cys (CGA)	30
Arg (UCU)	31
Phe (GAA)	32
Thr (UGU)	32
Val (UAC)	34
Gln (UUG)	35
Gly (UCC)	35
Asp (GUC)	36
Lys (UUU)	36
Ala (UGC)	37
Asn (GUU)	38
His (GUG)	38
Glu (UUC)	39
Ser (UGA)	39
Tyr (GUA)	40
Pro (UGG)	44
Average: 33	
Standard deviation: 6	

From Osawa *et al.* (1990).

the tRNA (or RF) gene are thereby reduced, allowing it to accumulate mutations that would be otherwise deleterious. If the codon is retired from use entirely, the corresponding tRNA (or RF) will no longer be maintained by selective forces and may be lost from the genome. Thus the codon will become unassigned.

From the above discussion, one may understand how and which codon is unassigned by directional mutation pressure. In short, an unassigned codon A may be produced only when synonymous codons exist. Synonymous codons, A and B, must be read by different anticodons, and the anticodon for B must not read A by wobbling. The anticodon for A becomes deleted upon conversion of A to B. Table 5.2 shows the effect of genomic G+C content (or directional mutation pressure) on tRNA anticodon compositions and codon–anticodon pairing patterns in eubacteria (Kano *et al.*, 1991). Stop codons are not included, and will be discussed in Section 6.1. All NNA codons may be unassigned by

GC-pressure, and CGG codons by AT-pressure. The nucleotide of the first position Leu and Arg codons may be silently substituted between a family box and a 2-codon set. As they are translated by different anticodons, CUN (Leu) and CGN (Arg) may be unassigned by AT-pressure and UUR (Leu) and AGR by GC-pressure (not shown in Table 5.1).

In mitochondria, unassigned codons (such as CGN in *Torulopsis glabrata*), would have been produced by directional mutation pressure according to the scheme described above. 'Genomic economization pressure' might also play a role. The size of mitochondrial genomes has decreased during evolution from simple to complex organisms by the discarding of many genes, including genes for tRNAs. In this process, certain tRNA genes seem to have degenerated gradually, and have finally been removed from the genomes (Osawa *et al.*, 1989*a*; Kurland, 1992). This constraint would have caused the corresponding codons to become unassigned. This process has not been well analysed (for further discussion of this see Section 6.6).

The principle for the production of unassigned codons in eukaryotes may be the same as that for bacteria, although no examples of unassigned codons have been reported. As noted in Section 2.2, the codon–anticodon pairing rules for eukaryotes would be somewhat different from those for bacteria, so that possible candidates for unassigned codons would differ. In eukaryotes: (1) anticodon INN in family boxes is said to read only codons NNC and NNU; and (2) anticodon *UNN is said to translate only codon NNA—codon NNG is read exclusively by anticodon CNN *in vivo*. If these rules are correct (and they may not be; see Section 2.2 and Suzuki *et al.*, 1994), then all codons except those for NNY in 2-codon sets, AUG (Met), and UGG (Trp) could be unassigned.

Table 5.3 summarizes the probable unassigned codons, and their causes, which have been reported.

5.2 Unassigned codons and stop codons

Stop codons are often referred to as nonsense codons because they do not correspond to any amino acid. However, the stop codons are not nonsensical; they function to release the synthesized peptides from the ribosomes by interacting with release factor. Unassigned codons also do not correspond to any amino acid. So how do stop codons and unassigned codons differ? As mentioned above, in *Mycoplasma capricolum*, codon CGG does not appear in coding frames (including termination sites) and $tRNA^{Arg}_{CCG}$ pairing with codon CGG is not detected. This suggests that an unassigned codon cannot function to terminate protein synthesis, even if it appears temporarily at the termination site. In a cell-free system prepared from *M. capricolum*, translation of synthetic mRNA containng in-frame CGG codons (Fig. 5.2) does not result in 'read-through' to codons (Tyr) beyond the CGG codons, i.e. translation ceases just before CGG. Sucrose-gradient centrifugation of the reaction mixture shows that the bulk of synthesized peptide is attached to 70S ribosomes (Fig. 5.3) and is released upon further incubation

Table 5.2 Effect of genomic G + C content on eubacterial tRNA anticodon

Bacterium (G + C content)	Family box: Codon	Family box: Anticodon	Family box (Arg CGN): Codon	Family box (Arg CGN): Anticodon
GC ↑				
Micrococcus luteus (75)	nnu	GNN	cgu	
	NNC		CGC	ICG
	(NNA)	–	cga	
	NNG	CNN	CGG	CCG
Escherichia coli (50)	NNU	GNN	CGU	
	NNC		CGC	ICG
	NNA	$^{+}$UNN	CGA	
	NNG	CNN	CGG	CCG
Bacillus subtilis (43)	NNU	GNN	CGU	
	NNC		CGC	ICG
	NNA	$^{+}$UNN	CGA	
	NNG		CGG	CCG
Mycoplasma capricolum (25)	NNU		CGU	
	nnc	UNN	cgc	ICG
	NNA		CGA	
AT ↓	nng		(CGG)	–

The table shows the general tendency of anticodon compositions in eubacteria with various genomic G + C contents and does not necessarily indicate that all the boxes and sets actually have these compositions. The bacteria are arranged in descending order of G + C content. Codons in small letters indicate those decreased in amount. Codons in parentheses are probable unassigned codons. –, deletion. See text for modifications of U, $^{+}$U and *U (from Kano *et al.* (1991), with modifications.

with puromycin. The result suggests that the peptide is in the P-site of the ribosome, in the form of peptidyl-tRNA, leaving the A-site empty. When in-frame CGG codons are replaced by UAA stop codons in mRNA, no read-through occurs beyond UAA, just as in the case of CGG. However, the synthesized peptide is released from 70S ribosomes, presumably by release factor 1 (RF-1) (Oba *et al.*, 1991*a*) (Fig. 5.4 (a–d)).

compositions and codon-anticodon pairing patterns

NNY 2-codon set		NNR 2-codon set		Ile AUY/A	
Codon	Anticodon	Codon	Anticodon	Codon	Anticodon
nnu	GNN			auu	GAU
NNC				AUC	
		(NNA)	-	(AUA)	-
		NNG	CNN		
NNU	GNN			AUU	GAU
NNC				AUC	
		NNA	*UNN	AUA	LAU
		NNG	CNN		
NNU	GNN			AUU	GAU
NNC				AUC	
		NNA	*UNN	AUA	LAU
		NNG			
NNU	GNN			AUU	GAU
nnc				auc	
		NNA	*UNN	AUA	LAU
		nng			

In *Micrococcus luteus*, certain codons ending with A do not appear in coding frames. The tRNAs with anticodon *UNN for most of these codons cannot be detected. *In vitro* translation experiments, similar to those of *Mycoplasma capricolum*, show that in-frame AGA and AUA, universal Arg, and Ile codons do not result in 'read-through' to the codon beyond AGA or AUA, and that the synthesized peptides are attached to 70S ribosomes. The in-frame stop codon UGA causes the release of peptides from the ribosomes (Kano *et al.*, 1993) (Fig. 5.5).

These data indicate that CGG in *Mycoplasma capricolum*, and AGA and AUA in *Micrococcus luteus*, are unassigned codons that differ from stop codons in that they are not used for termination, i.e. they are not recognized by release factors.

Table 5.3 Unassigned or nonsense codons

System	Probable unassigned codon	Cause
Mycoplasma capricolum	CGG (Arg)	AT-pressure; lack of $tRNA^{Arg}_{CCG}$
Micrococcus luteus	AGA (Arg)	GC-pressure; lack of $tRNA^{Arg}_{*UCU}$
	AUA (Ile)	GC-pressure; lack of $tRNA^{Ile}_{LAU}$
Torulopsis glabrata (a yeast) mitochondria	CGN (Arg)	AT-pressure; lack of $tRNA^{Arg}_{UCG}$
Prototheca wickerhamii (a green alga) mitochondria	CGG (Arg)	AT-pressure; lack of $tRNA^{Arg}_{CCG}$
	UGA (Stop)	AT-pressure; ? lack of RF-2
	or UAG (Stop)	AT-pressure; ? lack of RF-1

In the green alga, *Chlamydomonas reinhardtii* mitochondria, CGG (Arg), UGA (stop), and some other codons were not detected in 1887 codons examined (Boer and Gray, 1988).

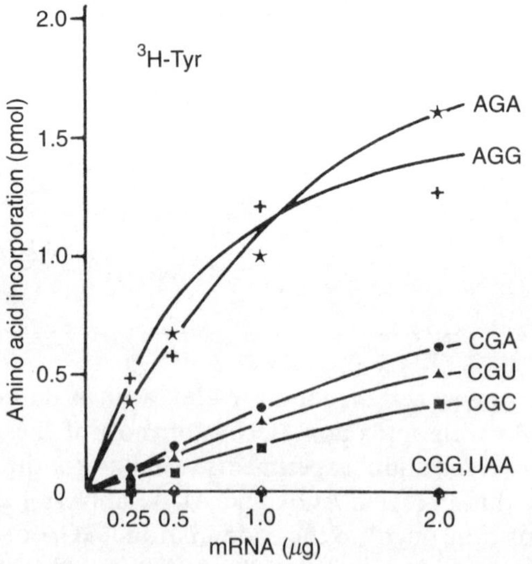

Fig. 5.2. Codon CGG is not translated in the *Mycoplasma capricolum* cell extract. Incorporation of [^{3}H]Tyr into peptides was measured with the *Mycoplasma capricolum* S30 fraction in a cell-free translation of various synthetic mRNAs. In-frame codons in the synthetic mRNA are shown by triplets. For the sequence of the synthetic mRNA, see Fig. 6.5 (from Oba *et al.*, 1991*a*).

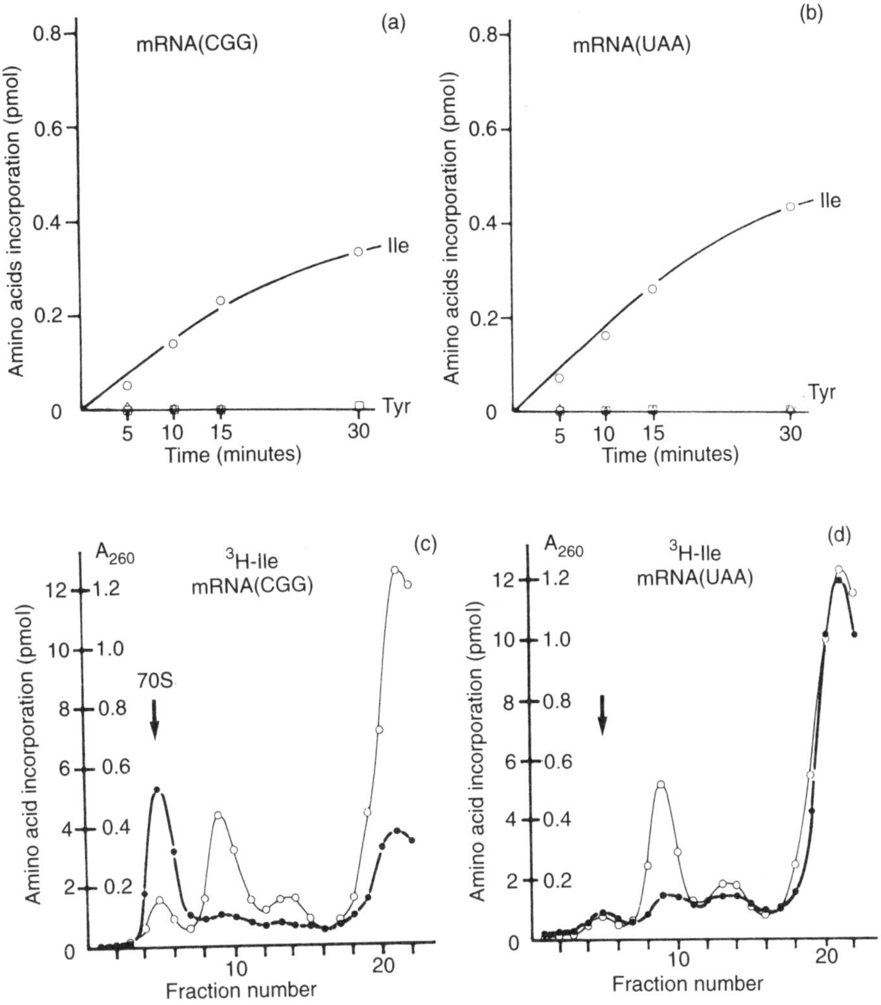

Fig. 5.3. Incorporation of [^{3}H]-labelled amino acids into peptides in a cell-free translation of synthetic mRNA (see Fig. 5.2) containing (a) codons CGG, and (b) codons UAA in the *Mycoplasma capricolum* S30 fraction. Sucrose-gradient centrifugation of reaction mixture labelled with [^{3}H]Ile. The incubated reaction mixture containing synthetic mRNA with (c) codons CGG or (d) codons UAA was centrifuged on a 5–20 per cent sucrose gradient and the radioactivity in the peptide fraction was measured (from Oba *et al.*, 1991*a*). For the sequence of mRNA, see Fig. 6.5.

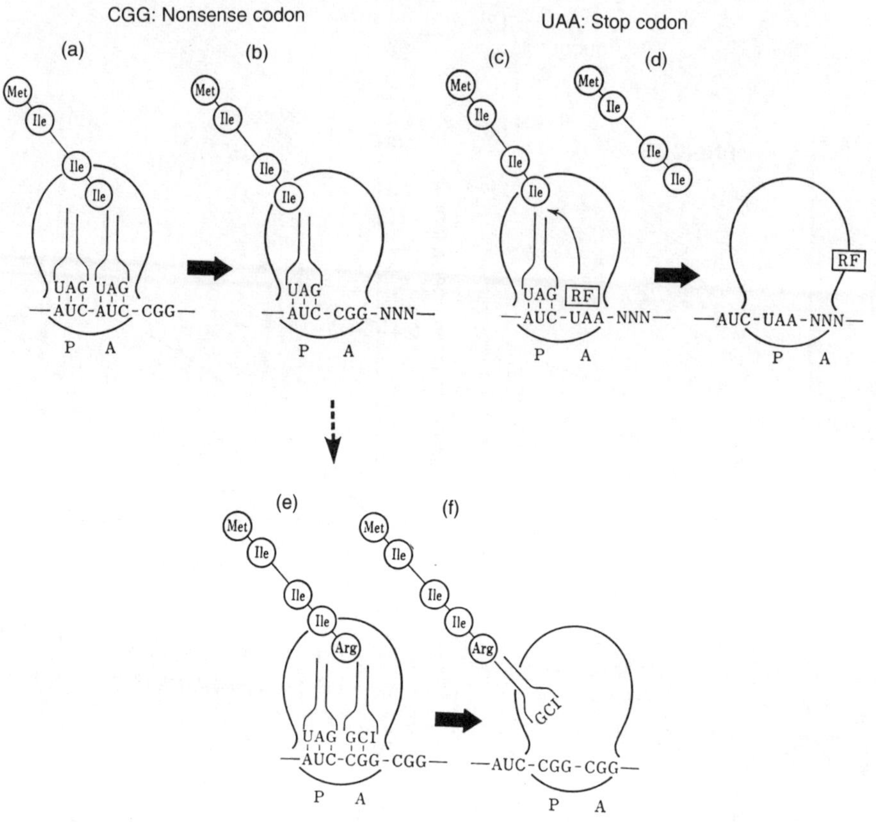

Fig. 5.4. Models for translation blockage by unassigned codon CGG (a and b). Compare (b) with (c) and (d), in which RF recognizes a stop codon UAA, followed by release of peptidyl-tRNA (d). Occasional false pairing of $tRNA^{Arg}_{ICG}$ with CCG codon (e) followed by release of peptidyl-tRNA (f) (from Oba *et al.*, 1991*a*, with modifications).

From the above description, it is evident that protein synthesis ceases at the site of the unassigned codon when an unassigned codon appears in the gene. The incomplete peptide remains attached to the ribosome, so that the ribosome cannot enter the next cycle of protein synthesis, i.e. it is inactivated. However, such an inactivation itself is not the main cause of negative selection, because it occurs in only a small fraction of ribosomes. The negative selection operates when an unassigned codon appears not to produce functional proteins. Note that the appearance of an unassigned codon, even at a non-essential site in a gene for functional proteins, is deleterious because the peptide cannot be released from ribosomes. This situation differs from that of stop codons, whose appearance in the coding frame at a site beyond the functional domain is often not deleterious, because truncated but functional peptide is released from the ribosome.

Other events can also take place. For example, when mRNA containing unassigned codon (e.g. CGG) was translated *in vitro* in the cell-free extracts of

```
-37                              S.D.     -1
5'GGGAACAAAAGCUGGGUACCGUUCUU AGGAG GAACUA

Met Ile Ile Ile Ile Phe Phe Phe Phe test codon Tyr Tyr Tyr Tyr Tyr Term
AUG AUC AUC AUC AUC UUC UUC UUC UUC  CGC  CGC  UAC UAC UAC UAC UAC UGA
                                     AGA  AGA
                                     AGG  AGG
                                     UGA  UGA
```

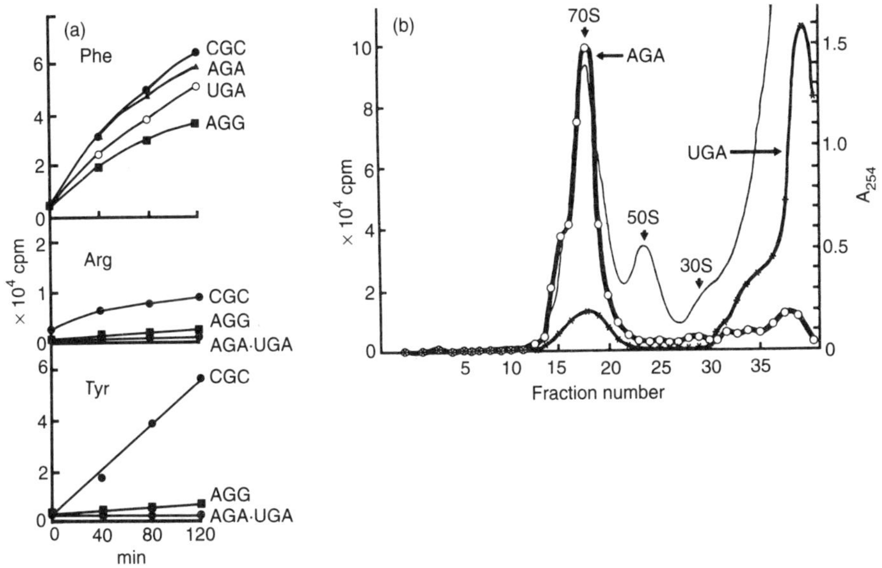

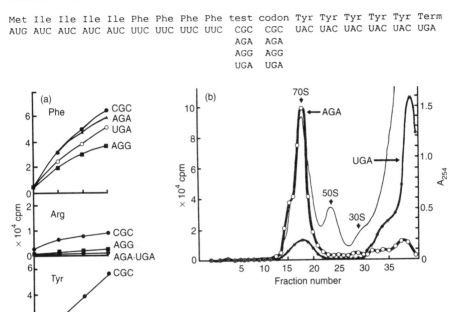

Fig. 5.5. AGA is probably an unassigned codon in *Micrococcus luteus*. (a) Incorporation of [^{3}H]Phe, [^{3}H]Arg and [^{3}H]Tyr into peptides was measured in cell-free translation of various synthetic mRNAs with the *Micrococcus luteus* S30 fraction. In-frame test codons (universal Arg codons CGC, AGG, AGA, or termination codon UGA) in the synthetic mRNA are shown by triplets. (b) Sucrose-gradient centrifugation of reaction mixture labelled with [^{3}H]Phe. The incubated reaction mixture containing synthetic mRNA with codons AGA (○) or UGA (×) was centrifuged on 5–20 per cent sucrose-gradient. The ultraviolet absorbance profiles for reaction mixture with mRNA (AGA) and with mRNA (UGA) were almost exactly the same, so that the two radioactive profiles are shown in one figure. The messenger RNAs used are shown above the figures. S.D., Shine–Dalgarno sequence. (From Kano *et al*., 1991.)

Mycoplasma capricolum, elongation of most of the peptides stopped before CGG codon was reached, leaving the A-site of the ribosome (CGG-site) empty, and the P-site occupied by the peptidyl-tRNA. As a result, transpeptidation would not occur. However, it is possible that another isoacceptor tRNA (in this case, arginyl-$tRNA^{Arg}_{ICG}$) entered the A-site accidentally by partial pairing of ICG with CGG (two-out-of-three!). As a result, the peptide of the peptidyl-tRNA on the P-site was transferred to arginyl-tRNA on the A-site (Fig. 5.4 (e, f)). Because I at the first anticodon position would not pair with G at the third codon position, pairing of codon CGG with anticodon ICG would not occur or, at best, would be unstable; thus the peptidyl-$tRNA^{Arg}_{ICG}$ would be released from the ribosome after transpeptidation. Indeed, some released material has been recognized and identified as peptidyl-$tRNA^{Arg}_{ICG}$ (Oba *et al.*, 1991*a*). This is further evidence against the two-out-of-three mechanism; the pairing of the third codon position with the first anticodon position would be essential for this to happen.

As discussed above, stop codons are not nonsense codons. It is therefore possible that a stop codon, like an amino acid codon, can become an unassigned codon by completely converting to another synonymous stop codon (e.g. UGA to UAA), along with the deletion or inactivation of the corresponding release factor (RF-2). The stop codon UAA that appears from mutation of UGA may be recognized by another release factor, RF-1, which does not interact with UGA. The whole process is therefore neutral. In this sense, release factors are equivalent to isoacceptor tRNAs (for a more detailed discussion of this see Section 6.1).

Contrary to the general textbook description that all organisms use the familiar genetic code table consisting of 64 codons, which includes three stop codons, the occurrence of unassigned codons implies that some life forms use fewer than 64 codons, although the genetic code is remarkably conserved among the majority of organisms. The number of usable codons may vary among organisms, and could decrease during evolution, e.g. by directional mutation pressure, or could increase up to 64 by capture of unassigned codons (Osawa *et al.*, 1990*b*; see Section 6.5).

6 The evolving genetic code

6.1 The frozen-accident theory

The 'universal' genetic code was established experimentally in around 1966, mainly by the use of *E. coli* systems. This code was thought to be common to all organisms and viruses, because it was the same in some very different organisms (e.g. yeast, vertebrates, and tobacco mosaic virus). Such apparent universality of the code led Crick to propose the frozen-accident theory. Crick (1968) states that: 'This accounts for the fact that the code does not change. To account for it being the same in all organisms one must assume that all life evolved from a single organism (more strictly, from a single closely interbreeding population)'. The theory further states that the proteins had become so sophisticated in a single pool of progenote cells that any changes in codon meaning would disrupt proteins by making unacceptable amino acid substitutions throughout their sequences, so that evolution of the code stopped, i.e. the code was frozen. 'In its extreme form, the theory implies that the allocation of codon to amino acids at this point was entirely a matter of chance' (Crick, 1968).

6.2 After the frozen-accident theory

In 1979, it was found that the genetic code in vertebrate mitochondria differed from the universal genetic code; it was found to use AUA for Met and UGA for Trp (Barrell *et al.*, 1979). Two new proposals were presented to account for this finding. The first was that, as the mitochondrial codon assignments are simplified relative to the universal code, they could represent a remnant of the primitive code that existed before the evolution of the universal code (Hasegawa and Miyata, 1980; Lewin, 1990). The second proposal was that, because the genomes of mitochondria are much smaller than the nuclear genomes, mitochondria can probably tolerate changes in the code that would not be acceptable to a larger and more complex system (Jukes, 1981; Lewin, 1990). This proposal is compatible with the frozen-accident theory. Indeed, the vertebrate mitochondrion synthesizes only some ten species of proteins. Meanwhile, a wide variety of species-specific changes in codon assignments has been reported for mitochondria from various organisms.

The first proposal cannot explain this diversity, unless the diffrent mitochondria were derived from 'primitive' bacteria of different origins. The second proposal is also unlikely, as codes that deviate from the universal genetic code were discovered in nuclear genomes in 1985. In *Mycoplasma capricolum*, UGA

Table 6.1 Variations in nuclear genetic code

Organism		UGA Stop	UAR Stop	CUG Leu	tRNA anticodon for altered code	Remarks
Mollicutes (Eubacteria)						
Mycoplasma	7 spp.	Trp	–	–	*UCA[a]	In *Acholeplasma laidlawii*, UGA is probably for stop
Spiroplasma	1 sp.	Trp	–	–	N	
Hemiascomycetes (yeasts)						
Candida	6 spp.	–	–	Ser	CAG[b]	In 8 *Candida* spp., CUG = Leu
Holotrichous ciliates						
Tetrahymena	2 spp.	–	Gln	–	UmUA for codons UAA and UAG; CUA for codon UAG[c]	
Paramecium	2 spp.	–	Gln	–	N	
Hypotrichous ciliates						
Stylonicia	1 sp.	–	Gln	–	N	
Oxytricha	2 spp.	–	Gln	–	N	
Euplotes	1 sp.	Cys	–	–	N	
Unicellular green algae						
Acetabularia	2 spp.	–	Gln	–	N	

– Same as universal code, or not determined; N, Not known; *U, cmnm^{5}Um.

a Determined for *Mycoplasma capricolum*. Gene for tRNA with anticodon UCA was found in *M. genitalium*, *M. pneumoniae*, and *M. gallisepticum*.

b Determined for *Candida cylindracea*, *C. parapsilosis*, *C. zeylanoides*, *C. albicans*, *C. rugosa*, and *C. melibiosica*.

c Determined for *Tetrahymena thermophilia*.

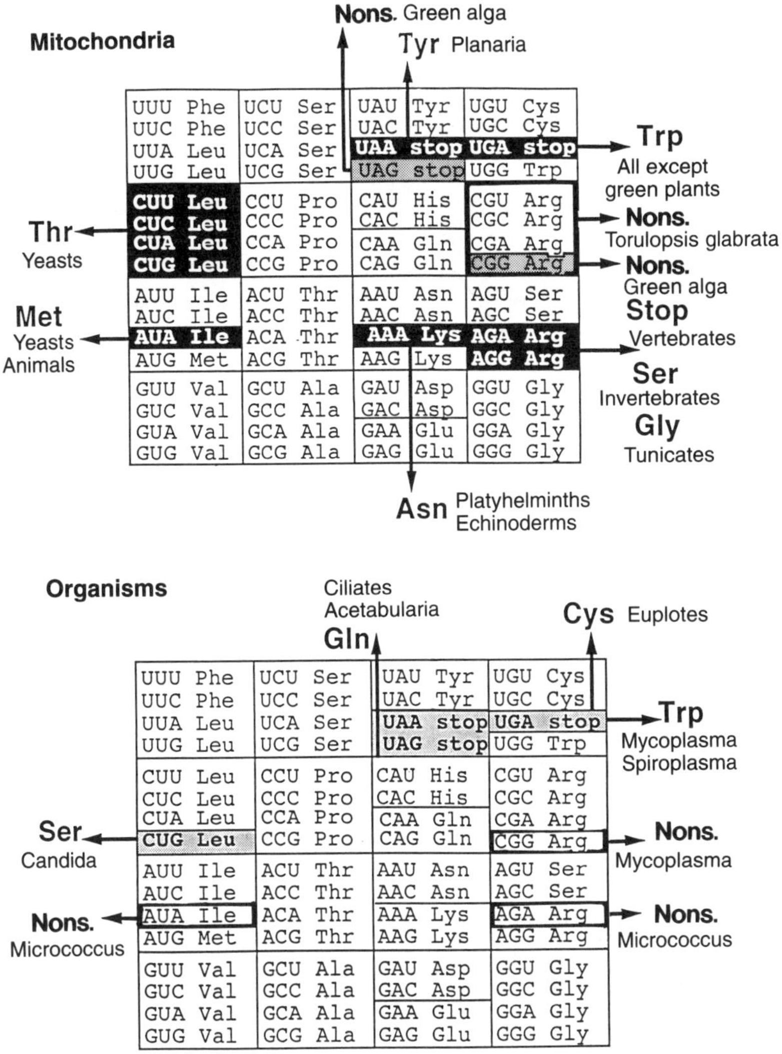

Fig. 6.1. Distribution of non-universal and unassigned codons in (a) mitochondria and (b) organisms. Nons. = Nonsense codon (based on Osawa *et al.*, 1990*b*).

codes for Trp (Yamao *et al.*, 1985) and, in certain ciliated protozoans, UAR codes for Gln (Caron and Meyer, 1985; Helftenbein, 1985; Horowitz and Gorovsky, 1985; Preer *et al.*, 1985). Some code changes in other nuclear genomes have also been reported. These facts, together with the occurrence of unassigned codons, indicate that the code is neither universal nor frozen. Rather, it is most likely that the code originated from the 'universal genetic code', which was used in single progenote cells. This code has evolved, and is still evolving, in both mitochondrial and nuclear genomes.

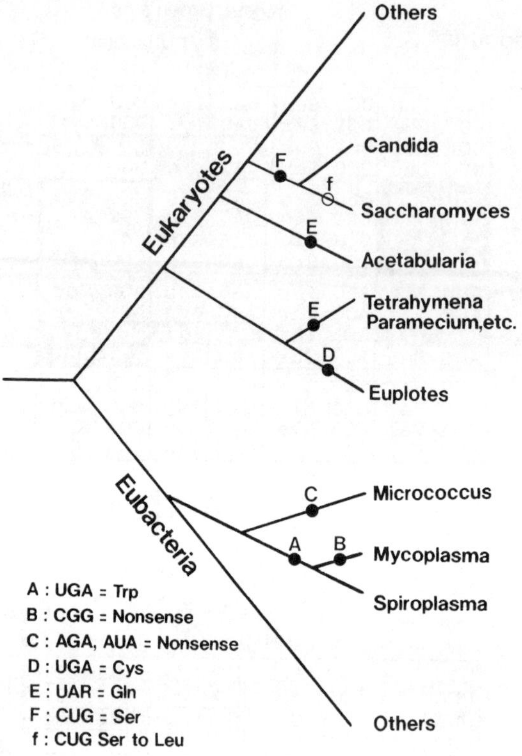

Fig. 6.2. Evolution of the genetic code in different organisms.

6.3 Distribution of non-universal codons and their translation

6.3.1 *The nuclear code*

The occurrence of universal and non-universal genetic codes in the nuclear systems of organisms is shown in Table 6.1 and Figs 6.1 and 6.2.

Mycoplasma capricolum and its related species use UGA as a Trp codon (Yamao *et al.*, 1985; Dudler *et al.*, 1988; Inamine *et al.*, 1988, 1990; Chevalier *et al.*, 1990; Kondo *et al.*, 1990; Renbaum *et al.*, 1990; Tham *et al.*, 1993). No altered codes have been found in other groups of eubacteria and archaebacteria (metabacteria). Evidence that, in addition to UGG, UGA is a Trp codon in *Mycoplasma/Spiroplasma* comes from the following:

1. The reading frames of genes for ribosomal and other proteins in these bacterial species contain many UGA Trp codons. Among them, a good number of UGA codons occur at the sites that are Trp in the corresponding *E. coli* proteins (Fig. 6.3).

2. tRNA$^{Trp}_{UCA}$, which can be charged with Trp and can translate both UGA and UGG by wobbling, has been found in several species of *Mycoplasma* (Fig. 6.4).
3. In-frame UGA codons in a synthetic mRNA are read as Trp as efficiently as UGG in a cell-free extract from *M. capricolum*, whereas only UGG (and not UGA) is translated in a similar system from *E. coli* (Oba *et al.*, 1991*b*) (Fig. 6.5).

```
    AA GCT TTA AAA GCT GGA GCT AAA GGA ATT AAA ACT GCT GTA AGT GGA AGA TTA GGT GGA GTT  062
M.c.   Ala Leu Lys Ala Gly Ala Lys Gly Ile Lys Thr Ala Val Ser Gly Arg Leu Gly Gly Val
E.c.   Ala Met Arg Leu Gly Ala Lys Gly Ile Lys Val Glu Val Ser Gly Arg Leu Gly Gly Ala

       GAA ATG GCA CGT ACT GAA GGA TAT TTA GAA GGT TCA GTA CCA CTA TCA ACT TTA AGA AAT  122
M.c.   Glu Met Ala Arg Thr Glu Gly Tyr Leu Glu Gly Ser Val Pro Leu Ser Thr Leu Arg Asn
E.c.   Glu Ile Ala Arg Thr Glu Trp Tyr Arg Gln Gly Arg Val Pro Leu His Thr Leu Arg Ala

       AAT ATT GAT TAT GCT TTA TAT GAA GCT CCA ACA ACA TAT GGT CAA ATT GGA GTT AAA GTA  182
M.c.   Asn Ile Asp Tyr Ala Leu Tyr Glu Ala Pro Thr Thr Tyr Gly Gln Ile Gly Val Lys Val
E.c.   Asp Ile Asp Tyr Asn Thr Ser Gln Ala His Thr Thr Tyr Gly Val Ile Gly Val Lys Val

        *
       TGA ATT AAT CAT GGT GAA GTA TTT --- --- --- --- AAA AAA GAA AGA ATG AAT AAT TCA  230
M.c.   TRP Ile Asn His Gly Glu Val Phe --- --- --- --- Lys Lys Glu Arg Met Asn Asn Ser
E.c.   Trp Ile Phe Lys Gly Glu Ile Leu Gly Gly Met Ala Ala Val Glu Gln Pro Glu Lys Pro
                                                           S3  End     L16 Start
       CAA ATA ATG GCA AAA CCA AGA ACT AAT AAA GGA GGT AAA AGA TAA TT- ATG TTA CAA CCA  289
M.c.   Gln Ile Met Ala Lys Pro Arg Thr Asn Lys Gly Gly Lys Arg ---     Met Leu Gln Pro
E.c.   Ala Ala Gln Pro Lys Lys Gln Gln Arg Lys Gly Arg Lys             Met Leu Gln Pro

       AAA AGA ACA AAA TAT CGT AAA CCT CAT AGA GTT AGT TAT GAA GGA AAA GCT AAA GGA GCT  349
M.c.   Lys Arg Asn Lys Tyr Arg Lys Pro His Arg Val Ser Tyr Glu Gly Lys Ala Lys Gly Ala
E.c.   Lys Arg Thr Lys Phe Arg Lys Met His Lys Gly Arg Asn Arg Gly Leu Ala Gln Gly Thr

                                                                        *
       AAA GAA ATT AAC TTT GGT GAA TTT GGT TTA ATG GCT TTA GAT GGT GCT TGA ATT GAT AAT  409
M.c.   Lys Glu Ile Asn Phe Gly Glu Phe Gly Leu Met Ala Leu Asp Gly Ala TRP Ile Asp Asn
E.c.   Asp --- Val Ser Phe Gly Ser Phe Gly Leu Lys Ala Val Gly Arg Gly Arg Leu Thr Ala

       CAT CAA ATA GAA GCT GCG CGT ATT GCT ATG ACA CGT TAT ATG AAG CGT GAT GGA AAA ATT  469
M.c.   His Gln Ile Glu Ala Ala Arg Ile Ala Met Thr Arg Tyr Met Lys Arg Asp Gly Lys Ile
E.c.   Arg Gln Ile Glu Ala Ala Arg Arg Ala Met Thr Arg Ala Val Lys Arg Gln Gly Lys Ile

        *
       TGA ATG AGA ATT TTC CCA CAT ATG GCA ATG ACT AAA AAA CCT GCT GAA GTT CGT ATG GGT  529
M.c.   TRP Met Arg Ile Phe Pro His Met Ala Met Thr Lys Lys Pro Ala Glu Val Arg Met Gly
E.c.   Trp Ile Arg Val Phe Pro Asp Lys Pro Ile Thr Glu Lys Pro Leu Ala Val Arg Met Gly

                                        *
       TCA GGA AAA GGA AAT CCT GAA AAA TGA GTA GCA GTA GTT AAA AAA GGA ACA ATT ATG TTT  589
M.c.   Ser Gly Lys Gly Asn Pro Glu Lys TRP Val Ala Val Val Lys Lys Gly Thr Ile Met Phe
E.c.   Lys Gly Lys Gly Asn Val Glu Tyr Trp Val Ala Leu Ile Gln Pro Gly Lys Val Leu Tyr

       GAA GTT GCT CAA GTA AAT GAG CAA GTA GCT AGA GAA GCT T                            629
M.c.   Glu Val Ala Gln Val Asn Glu Gln Val Ala Arg Glu Ala
E.c.   Glu Met Asp Gly Val Pro Glu Glu Leu Ala Arg Glu Ala
```

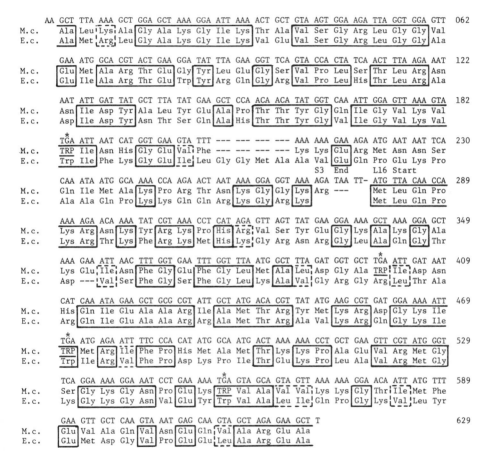

Fig. 6.3. DNA sequence of a part of *Mycoplasma capricolum* S3 and L16 ribosomal protein genes. The DNA sequence of the mRNA-like strand, together with the predicted amino acid sequence of *M. capricolum* ribosomal protein S3 and L16 genes (M.c.) was aligned with the corresponding *E. coli* protein sequence (E.c.). The sites for identical amino acids and those of conservative amino acid substitution are boxed with solid and dotted lines, respectively. TGA triplets are marked with asterisks (from Yamao *et al.*, 1985).

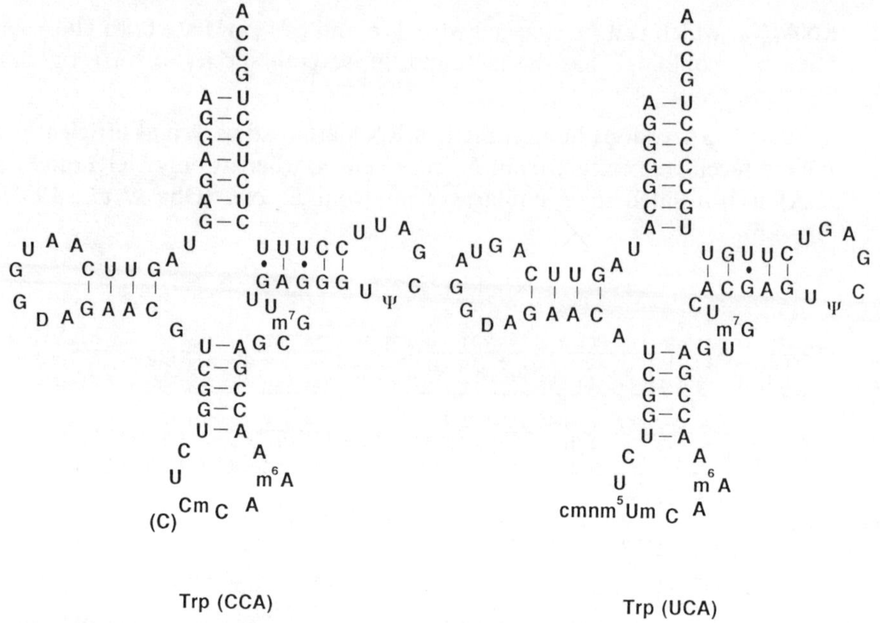

Fig. 6.4. Secondary structures of *Mycoplasma capricolum* $tRNA^{Trp}_{CCA}$ and $tRNA^{Trp}_{UCA}$. $tRNA^{Trp}_{CCA}$ translates codon UGG, and $tRNA^{Trp}_{UCA}$ translates both UGA and UGG (from Andachi *et al.*, 1989).

```
pppGGGCGAAUUGGAGCUC AGGAGG GUUAC
                    <S.D.>

AUG AUCAUCAUCAUCAUCAUCAUCAUUAUUAUU UGAUGA UACUACUACUACUAC UAA
Met <----------10 x Ile---------->        <---5 x Tyr---> Stop

                                              CUGCAGGAAUUCGAUAUCAAGCU
```

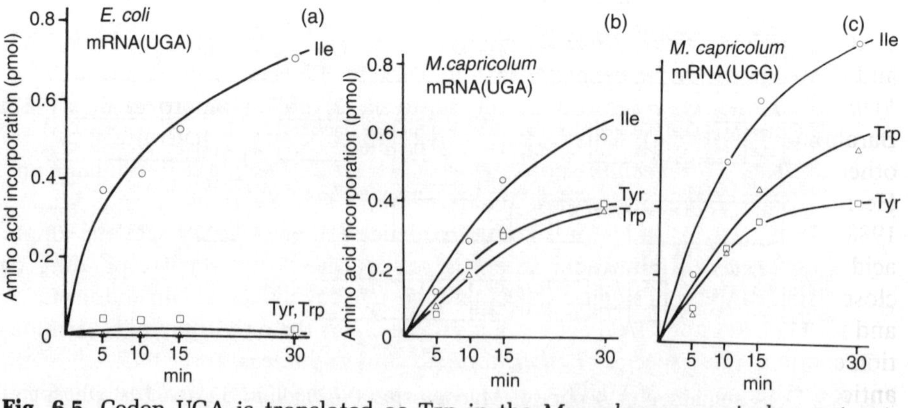

Fig. 6.5. Codon UGA is translated as Trp in the *Mycoplasma capricolum* extract. Incorporation of [^{3}H]amino acids into peptides was measured in the cell-free translation system of (a) *Escherichia coli* and (b and c) *Mycoplasma capricolum*. The mRNAs used are shown above the figures. S.D., Shine–Dalgarno sequence; 10 × Ile, 10 Ile codons; 5 × Tyr, 5 Tyr codons. In (c), two UGAs are replaced by UGGs (from Oba *et al.*, 1991*a*).

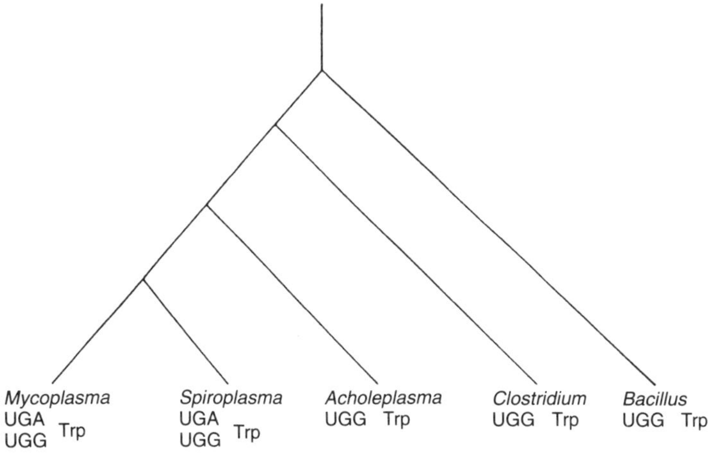

Fig. 6.6. Branching order of several Gram-positive eubacteria and their Trp codon usage (from Osawa *et al.*, 1992).

Figure 6.6 shows a phylogenetic tree of the Gram-positive eubacterial group constructed from the 5S rRNA sequence comparisons. *Acholeplasma laidlawii*, a member of the class *Mollicutes* (which also contains the genera *Mycoplasma* and *Spiroplasma*) does not use UGA for Trp (Tanaka *et al.*, 1989). The code change UGA for Trp evidently occurred in the *Mycoplasma/Spiroplasma* lineage after its separation from *Acholeplasma laidlawii* and from their common ancestor—and occurred not more than 0.6 billion years ago according to the phylogenetic tree of 5S rRNA (Osawa *et al.*, 1992).

Several ciliated protozoans possess two universal stop codons—UAA and UAG—which code for Gln (CAA and CAG also code for Gln). Codons UAA and UAG appear in the genes for histone H3 and H2, and for tubuline in *Tetrahymena thermophilia* (Horowitz and Gorovsky, 1985; Nomoto *et al.*, 1987; Barahona *et al.*, 1988). This internal presentation occurs in several genes from other ciliates—*Paramecium* spp., *Oxytricha* (Herrick *et al.*, 1987; Williams and Herrick, 1991), and *Stylonichia* (Helftenbein, 1985; Helftenbein and Müller, 1988). Comparisons of the nucleotide sequences of these genes with the amino acid sequences or amino acid compositions of the proteins corresponding or closely related to them suggest that, in addition to the universal Gln codons CAA and CAG, UAA and UAG are read as Gln. UGA is the only functional termination codon in these species. *T. thermophilia* contains at least three $tNRAs^{Gln}$ with anticodon sequences of UUG (for regular Gln codons CAR), UmUA, and CUA (Fig. 6.7) (Kuchino *et al.*, 1985; Hanyu *et al.*, 1986). The $tRNA^{Gln}_{UmUA}$ reads codon UAA in α-globulin mRNA and UAG of tobacco mosaic virus RNA, whereas $tRNA^{Gln}_{CUA}$ recognizes only UAG (Hanyu *et al.*, 1986).

Another ciliate—*Euplotes octocarinatus*—uses UGA as a Cys codon (Meyer *et al.*, 1991) and UAA as a stop codon (Harper and Jahn, 1989). In this ciliate,

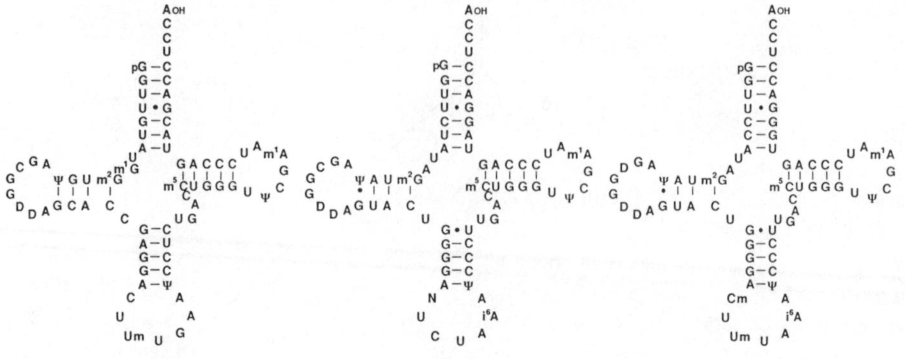

(a) $tRNA^{Gln}_{(UmUG)}$ (b) $tRNA^{Gln}_{(CUA)}$ (c) $tRNA^{Gln}_{(UmUA)}$

Fig. 6.7. Secondary structures of three species of *Tetrahymena thermophilia* $tRNA^{Gln}$. $tRNA^{Gln}_{UUG,\ CUA,\ and\ UUA}$ are responsible for translation of codons CAR, UAG, and UAR (mainly UAA), respectively (from Hanyu *et al.*, 1986).

codon UGA appears in the pheromone 3 gene at the Cys sites of this protein (Meyer *et al.*, 1991).

In protozoology, *Tetrahymena* and *Paramecium* spp. are classified as holotrichous ciliates, whereas *Euplotes*, *Oxytricha*, and *Stylonichia* belong to the hypotrichous ciliates. The molecular phylogenetic tree supports this view (Hendriks *et al.*, 1988); *Euplotes* branched off earlier than *Oxytricha/Stylonichia*. In view of the reported changes, one must suppose that the ancestor leading to these two ciliate groups used UAR as stop codons and that the change in UAR from stop codons to Gln occurred independently—twice—once in the holotrichous line and once in the ancestor of *Oxytricha/Stylonichia* after *Euplotes* had branched off. Another change—UGA to Cys—took place in *Euplotes*. However, further investigations, such as identification of the responsible tRNA, are needed to establish the assignment of UAR codons in *Oxytricha/Stylonichia/Euplotes*.

Interestingly, two *Acetabularia* species (unicellular green algae which, phylogenetically, are a long way from ciliates) also use UAR for Gln codons, as was found in cDNA clones of the small subunit of ribulose-1,5-bisphosphate carboxylase gene (Schneider *et al.*, 1989). The tRNA responsible for these codons has not been identified.

The phylogenetic position of ciliates is near the animal kingdom in both the 5S rRNA tree (Hori and Osawa, 1987) and the classical tree, although some other molecular trees show ciliates as branching off somewhat earlier. However, any branching point must be after prokaryotic emergence. In the prokaryotic genetic code, codons UAR are used as stops and UGA is used as stop or, rarely, as Trp. Furthermore, the change UAR to Gln from stops also occurred independently in green algae—*Acetabularia* spp. Differences in nucleotide sequences between the new tRNAs that translate UAR codons, and the other $tRNA^{Gln}$ in *Tetrahymena* are—in $tRNA^{Gln}_{UmUG}$—13 nucleotide substitutions compared to $tRNA^{Gln}_{UmUA}$ and 11 nucleotide substitutions compared to $tRNA^{Gln}_{CUA}$. In $tRNA^{Gln}_{UmUA}$ the difference is 4 nucleotide substitutions compared to $tRNA^{Gln}_{CUA}$. These differences may be a rough clue to the time since evolutionary divergence. Differences between yeast and rat or wheat tRNAs with the same anticodon are 13 to 15 substitutions, and differences between *Drosophila* and vertebrates are 5 to 12 substitutions. From these facts, one must conclude that this code change—stop to Gln—took place since the evolution of the eukaryotes (Jukes *et al.*, 1987).

CUG, a universal Leu codon, is assigned to Ser in the yeast *Candida cylindracea* (Kawaguchi *et al.*, 1989) (Fig. 6.8) and its related species (Ohama *et al.*, 1993). Amongst the nuclear genetic code of eukaryotes and prokaryotes, this is a unique instance in which the assignment of an amino acid codon deviates from the universal code. The evidence for this non-universal assignment of the codon CUG is derived from:

1. CUG is the major Ser codon in the lipase gene of *C. cylindracea* (Kawaguchi *et al.*, 1989).
2. Translation of synthetic mRNAs with CUG codons in-frame in the cell-free extracts indicates that, in six *Candida* spp. (including *C. cylindracea*) CUG is read as Ser, whereas in eight other hemiascomycetes-species (including *Saccharomyces cerevisiae*) CUG is translated as Leu (Ohama *et al.*, 1993) (Fig. 6.9).
3. $tRNA^{Ser}$ with anticodon sequence CAG, which is complementary to codon CUG, has been found in all six *Candida* species in which CUG is used for a Ser codon (Fig. 6.10). The $tRNA^{Ser}_{CAG}$ gene of two *Candida* spp. (*C. cylindracea* and *C. melibiosica*) are interrupted by an intron (Fig. 6.11). The $tRNA^{Ser}_{CAG}$ can be derived directly from what was previously $tRNA^{Ser}_{IGA}$, because there is a C insertion in the intron-containing tRNA gene to create anticodon CAG by splicing (Fig. 6.12). The $tRNA^{Ser}_{CAG}$ gene of the four other species lacks intron (Yokogawa *et al.*, 1992; Ohama *et al.*, 1993). Presumably, the intron has disappeared during evolution of the $tRNA^{Ser}_{CAG}$ gene.

Distribution of this non-universal genetic code in various yeasts has been studied on the basis of their phylogenetic relationships by 5S rRNA sequence comparisons (Hori, H. *et al.*, unpublished data).

The phylogenetic relationships of these yeasts suggest that the assignment of codon CUG has changed twice—from Leu to Ser, and then as a reversal to Leu again—during the evolution of yeasts (Fig. 6.13). *Candida* spp. in which CUG is a Ser codon, branched off relatively recently from the euascomycetes (*Aspergillus*, *Pencillium*, and *Neurospora*) and share a common ancestry with hemiascomycetes, in all of which the codon CUG is for Leu and not for Ser. Branching of these *Candida* species from other species of hemiascomycetes, such as *Saccharomyces* and *Torulopsis* (CUG = Leu) seems to be an even more recent event. The changes of CUG to Ser from Leu, and its reversal, would have occurred nearly

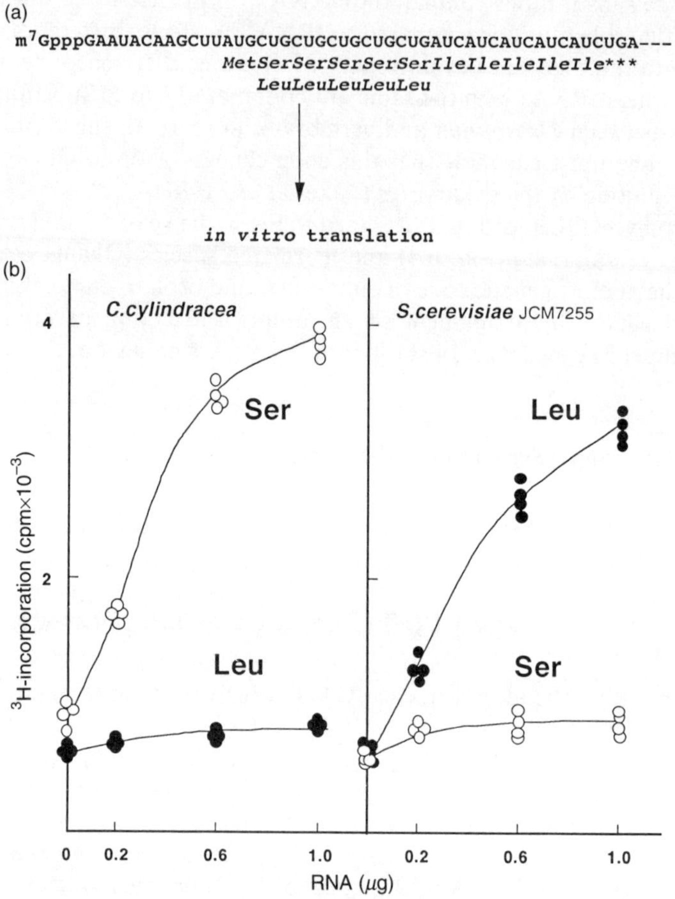

Fig. 6.8. Codon CUG is read as Ser in *Candida cylindracea*. Incorporation of [^{3}H]amino acids into peptides was measured in the cell-free translation system of *Candida cylindracea* and *Saccharomyces cerevisiae*. The mRNA used is shown above the figure (from Kawaguchi *et al.*, 1989).

0.9 and 0.15 billion years ago, respectively, according to the 5S rRNA phylogenetic tree (Hori and Osawa, 1987).

When non-universal genetic codes were discovered in *Mycoplasma capricolum* and in ciliated protozoans, Grivell (1986) proposed that these codes may be 'relics of a primitive genetic code that possibly preceded the emergence of the so-called universal code', and that 'the ciliates branched off from the primitive eukaryotic ancestor very early in evolution, quite possibly at a time when protein translation was still tolerant of changes in the genetic code'. Lewin (1990) gave a similar explanation:

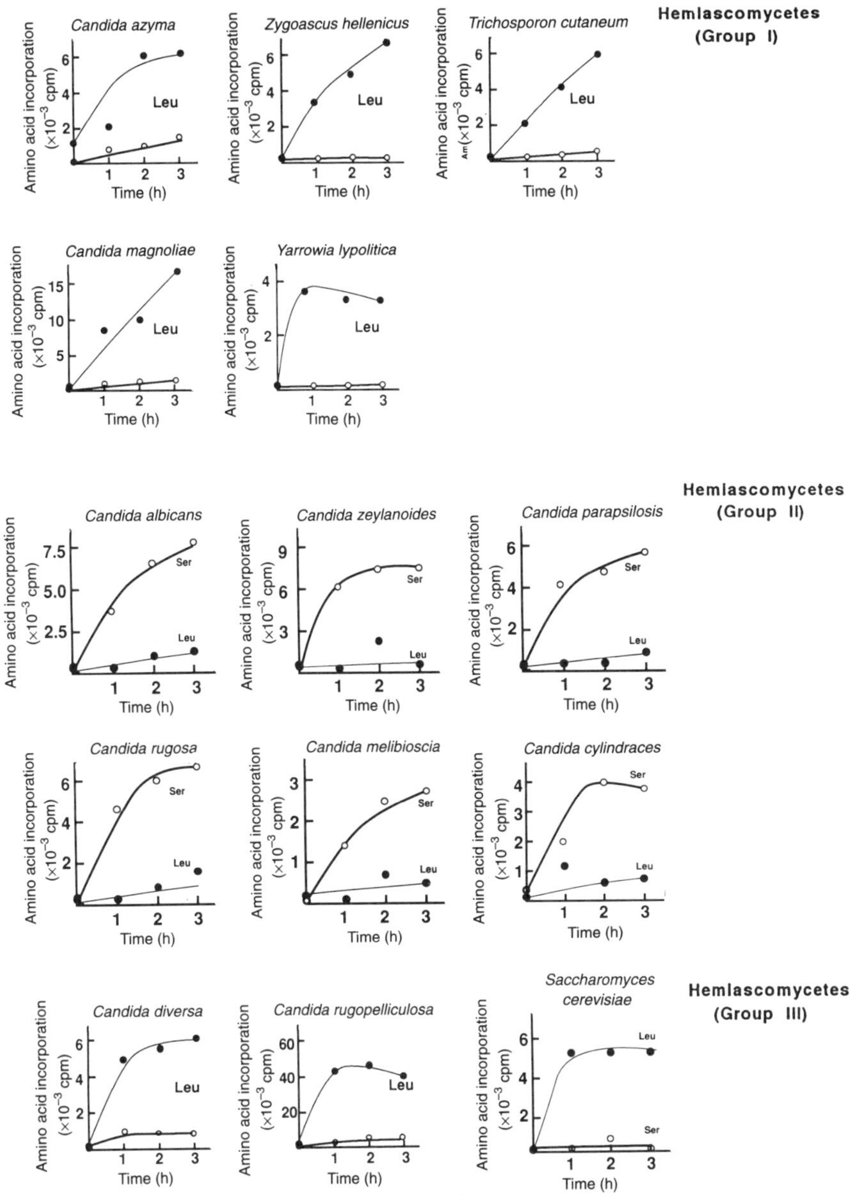


```
Kpn I
1         10             20             30             40             50
TGGGTACCAA AACACCAAGA ACTTAGTTTC GAATAAACAC ACATAAATAA ACAAA

                                                                       Hind III
     60             70             80             90             100
ATG ATG CTG CTG CTG CTG CTG CTG TTC TTC TTC TTC TTC TTC TTC TAA AAGCTTGA
Met Met Leu Leu Leu Leu Leu Leu Phe Phe Phe Phe Phe Phe Phe ***
        /Ser/Ser/Ser/Ser/Ser/Ser
```

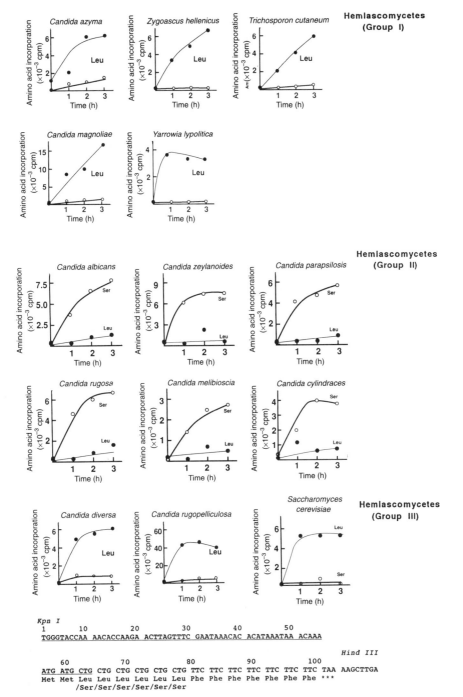

Fig. 6.9. *In vitro* translation of synthetic messenger RNA with the S30 fractions from various *Hemiascomycetes* (yeasts) spp. The mRNA used is shown below the figure. Codon CUG is read as Leu in two groups of yeasts (Group I and Group III), and as Ser in several *Candida* spp. (Group II) (from Ohama *et al.*, 1993).

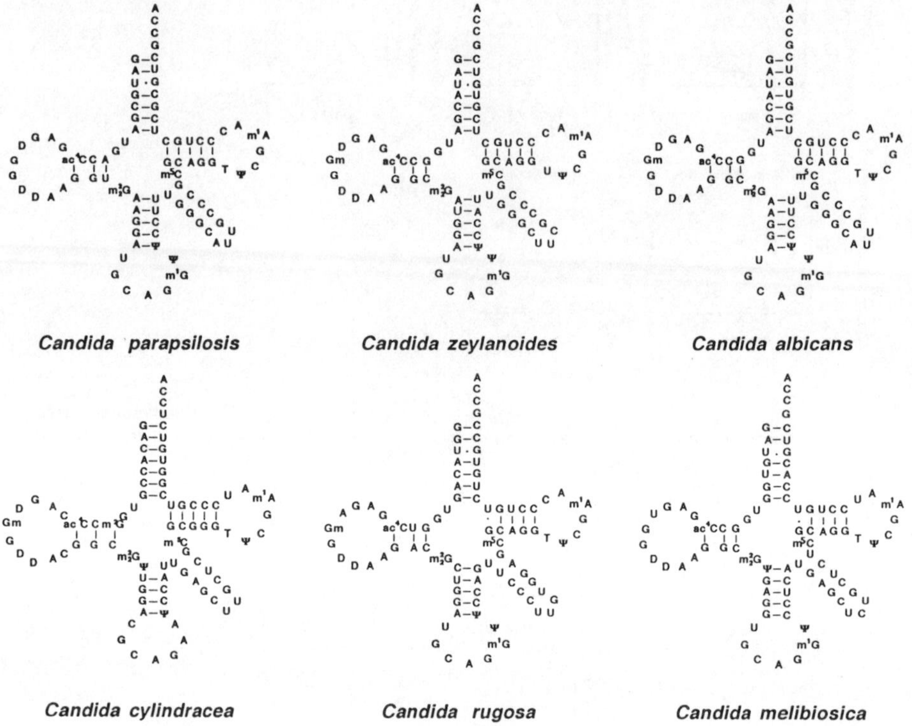

Fig. 6.10. Secondary structures of $tRNA^{Ser}_{CAG}$ responsible to translate codon CUG as Ser in various *Candida* spp. Note that the anticodon sequence of all these tRNAs is CAG, which can read codon CUG (from Ohama *et al.*, 1993).

At this time (when ciliates branched) the code was still quite error-prone, with termination in particular being an erratic process. Perhaps in fact termination was so leaky that many proteins ended (if not at random sites) at any one of many possible sites. In the major branch of evolution, termination became effective at all the termination codons as the precision of protein synthesis increased, but in *Mycoplasma* and ciliates only some of the termination codons were retained.

Meyer *et al.* (1991) gave a similar explanation for the change of codon UGA from stop to Cys in *Euplotes*. The proposals that 'translation was tolerant of changes' and that translation was 'an erratic process' have no basis in fact. An entirely different explanation is more probable—these altered codes may be of recent origin as discussed above, and derived from the universal genetic code.

6.3.2 *The mitochondrial code*

Mitochondria can be divided into two categories—plants and non-plants—with respect to code usage. Plant mitochondria use the universal genetic code, whereas

Plate 1. A plaque showing the creation of the animals. Southern Italian, *c.* 1080 (copy obtained from the Metropolitan Museum of Art, NY). The Creator and all these creatures use the common 'universal' genetic code. Photo by Mrs M. Motoyama, Biohistory Research Hall.

The organisms illustrated in Plates 2 to 15, and all mitochondria except those in green plants and a few others, use deviant genetic codes.

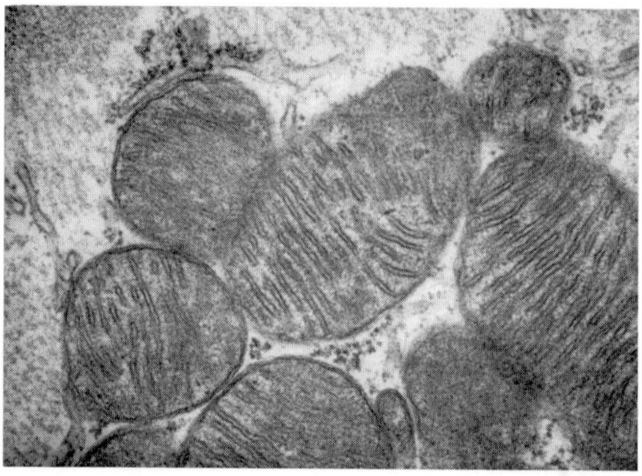

Plate 2. Rat liver mitochondria (TEM; Courtesy of Dr T. Kuroiwa, University of Tokyo).

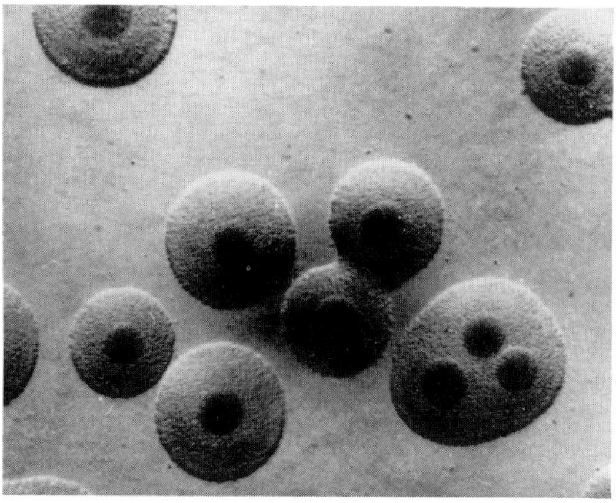

Plate 3. *Mycoplasma capricolum* (colonies) (Courtesy of Dr R. Harasawa, University of Tokyo).

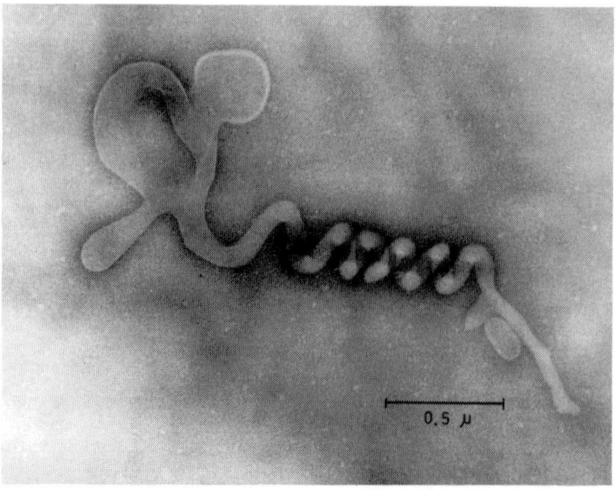

Plate 4. *Spiroplasma kunkelii* var. *allistephi* (TEM; Courtesy of Dr F. Kondo, Miyazaki University).

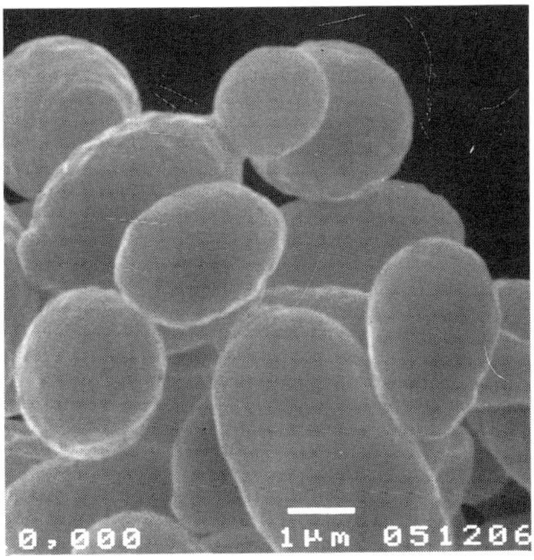

Plate 5. *Candida parapsilosis.*

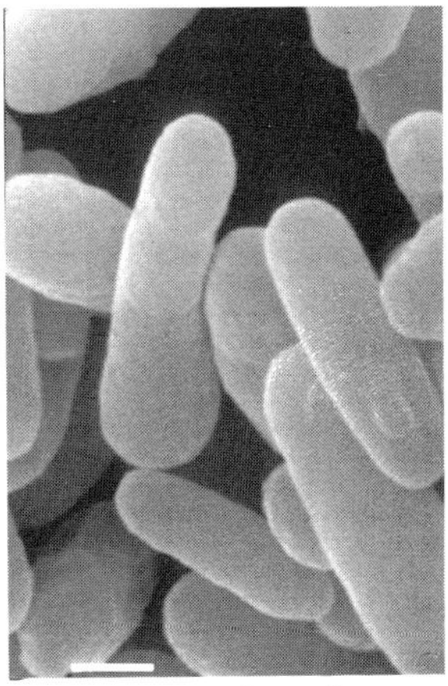

Plate 6. *C. zeylanoides.*

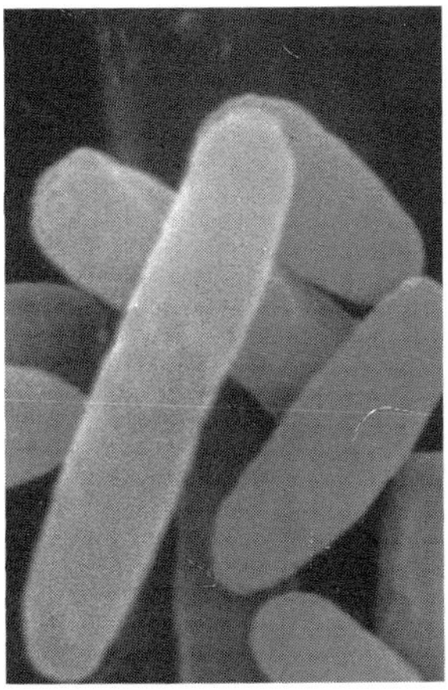

Plate 7. *C. cylindracea.*

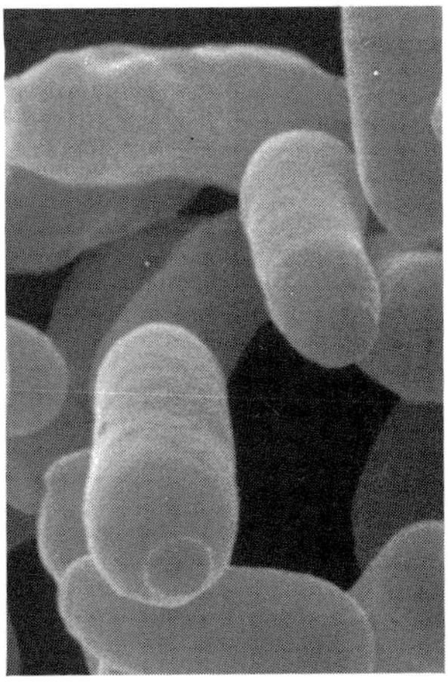

Plate 8. *C. rugosa.*

Plate 9. *C. melibiosica* (Plates 4–8, SEM; Courtesy of Dr T. Ueda, University of Tokyo, and Dr M. Yoshida and Mrs M. Motoyama, Biohistory Research Hall).

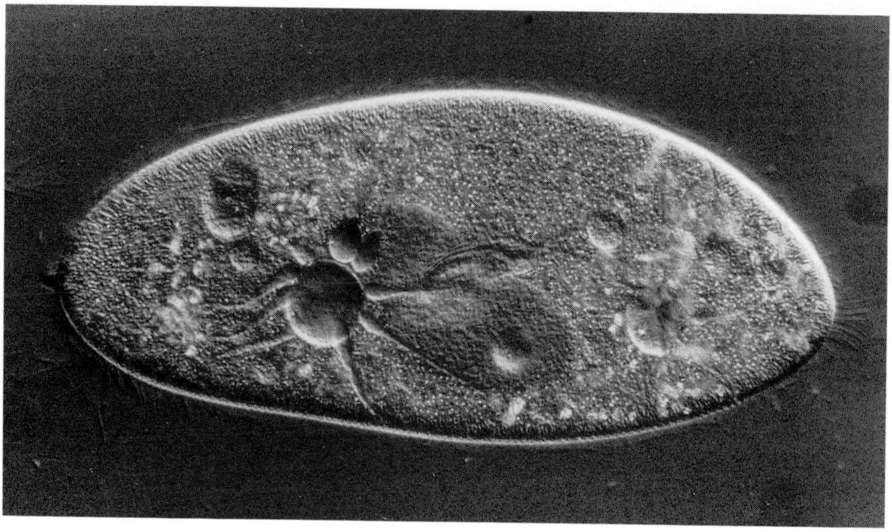

Plate 10a. *Paramecium caudatum* (SEM; Courtesy of Dr T. Kosaka, Hiroshima University).

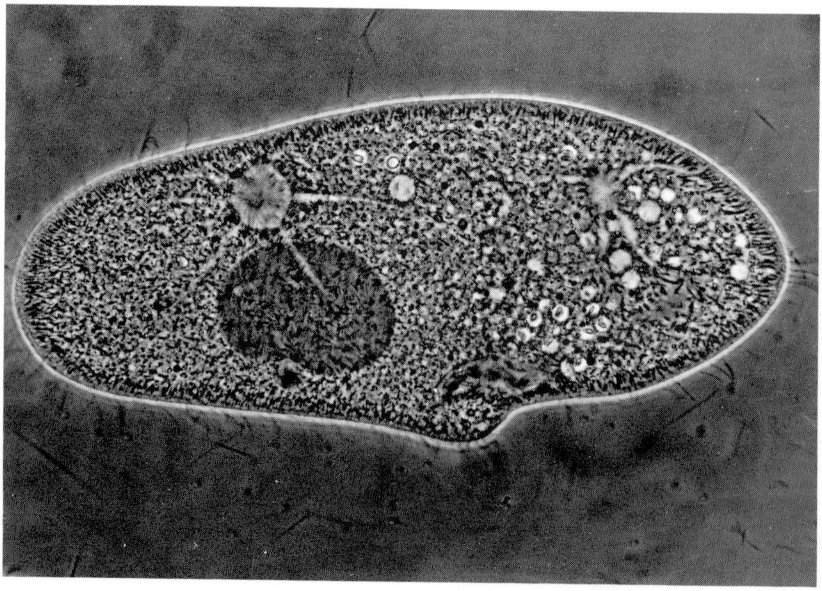

Plate 10b. *Paramecium caudatum* (Courtesy of Dr K. Hiwatashi, Ishinomaki Senshu University).

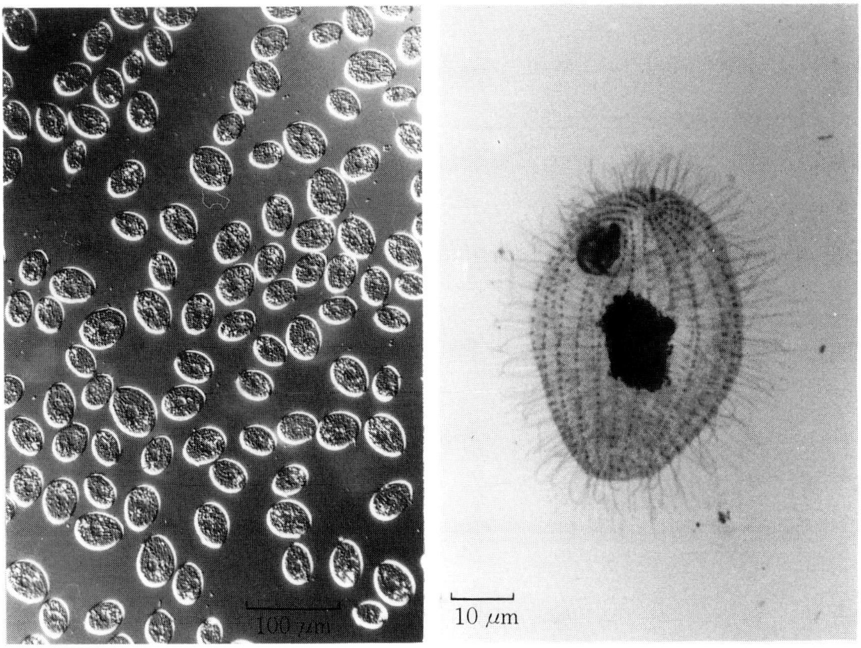

Plate 11. *Tetrahymena thermophilia*.(Courtesy of Dr K. Hiwatashi).

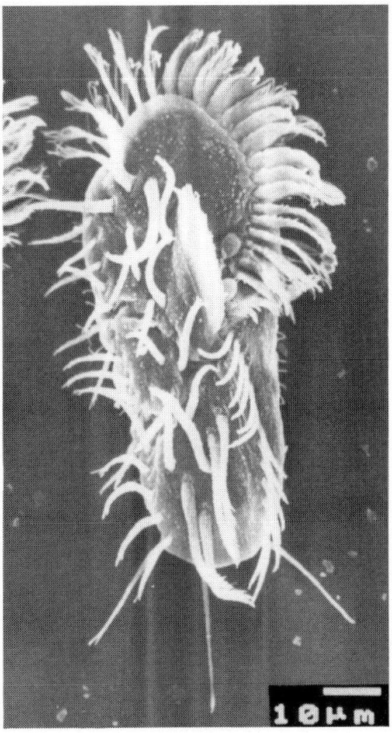

Plate 12. *Stylonichia lemnae* (SEM; Courtesy of Dr T. Takahashi, Hiroshima University).

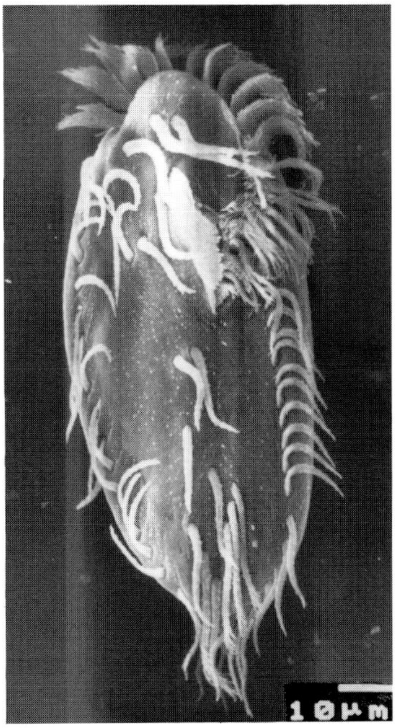

Plate 13. *Oxytricha* sp. (SEM; Courtesy of Dr T. Takahashi).

Plate 14. *Euplotes woodrufii* (Courtesy of Dr T. Kosaka).

Plate 15. *Acetabularia* sp. (Photo © I. Soyama).

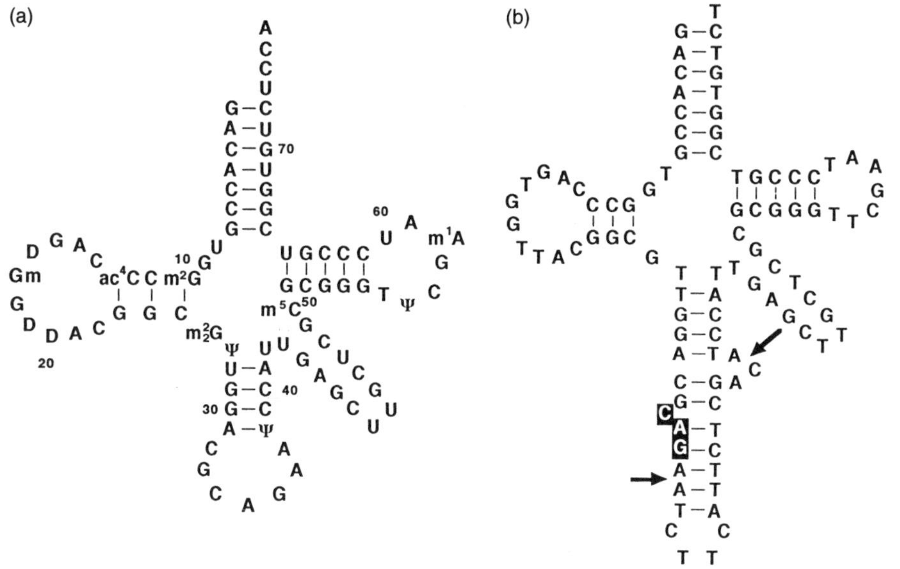

Fig. 6.11. Secondary structures of (a) $tRNA^{Ser}_{CAG}$ and (b) its gene in *Candida cylindracea*. The anticodon is shown by a solid background; → splicing sites for the tRNA gene (from Yokogawa *et al.*, 1992).

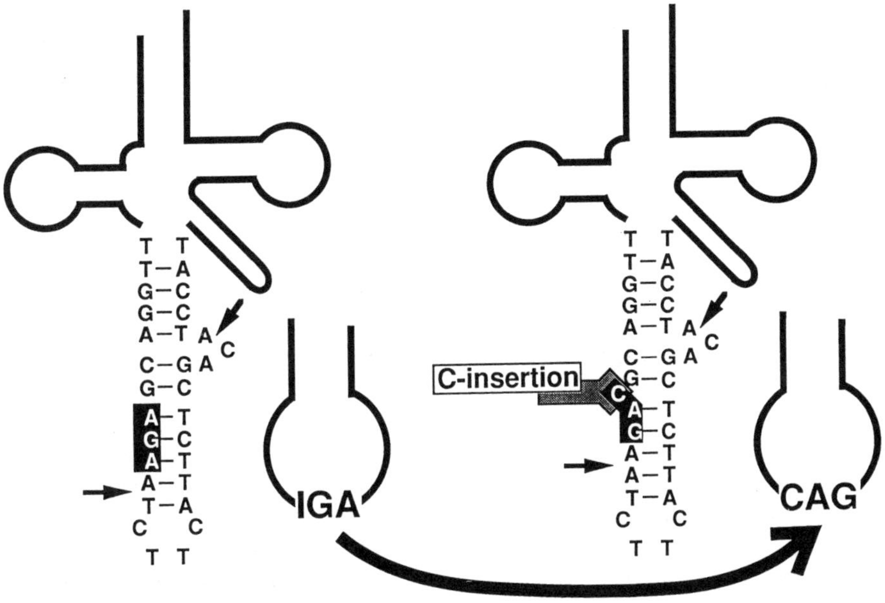

Fig. 6.12. Creation of $tRNA^{Ser}_{CAG}$ gene from $tRNA^{Ser}_{IGA}$ gene by splicing. The anticodon is shown by a solid background. Splicing sites for the $tRNA^{Ser}_{CAG}$ gene and possible splicing sites for the putative ancestral $tRNA^{Ser}_{IGA}$ gene, to which cytidine insertion at the position indicated would have occurred, are indicated by solid arrows (drawn from data from Yokobori *et al.* (1993); courtesy of K. Watanabe).

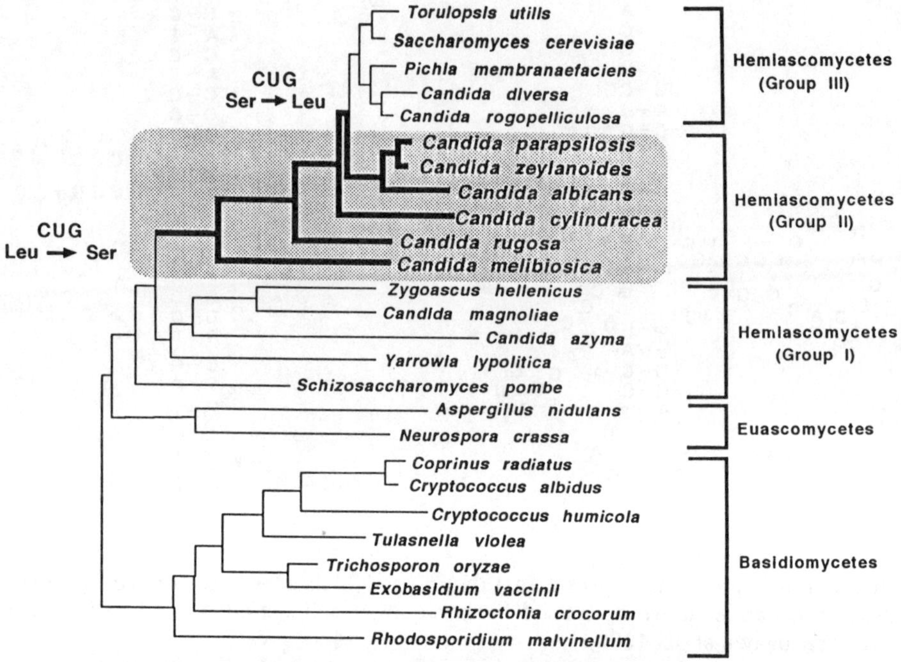

Fig. 6.13. The phylogenetic tree of yeasts and related species based on 5S rRNA sequences with amino acid assignment of codon CUG. Species in the shadowed area use CUG for Ser; in other species CUG = Leu (courtesy of K. Watanabe).

non-plant mitochondria have many code changes. The code variations were first found in 1979 by comparison of the nucleotide sequences of the human COII (cytochrome oxidase subunit II) gene with the amino acid sequences of bovine COII. This led to the conclusion that UGA and AUA specify Trp and Met, respectively (Barrell *et al.*, 1979). At about the same time, yeast mitochondrial genes for cytochrome *b*, COII, and ATPase oli 2 were found to use UGA as a Trp codon (Macino *et al.*, 1979).

Since then, the non-universal mitochondrial genetic code has been deduced mostly by comparing the DNA sequences with the corresponding sequences of other mitochondrial DNA in which each codon had been already assigned (e.g. de Bruijn, 1983; Himeno *et al.*, 1987). Thus, although most of the reported code changes are probably correct, they are by no means conclusive because of, for example, RNA editing in certain systems (see Sections 8.2 to 8.6).

As a result of genomic economization, non-plant mitochondria contain a limited set of tRNAs (22 species in metazoan mitochondria and 24 in yeast mitochondria). Most metazoan mitochondrial tRNAs have unusual structures, lacking the D-loop–T-loop interaction (Anderson *et al.*, 1981, 1982). Cytoplasmic tRNAs are not imported into animal mitochondria (Roe *et al.*, 1981). Certain features characteristic of mitochondrial tRNAs are largely responsible for reading non-universal codons in mitochondria.

Table 6.2 Variations in mitochondrial genetic code

Organism	UGA Stop	AUA Ile	AAA Lys	AGR Arg	CUN Leu	UAA Stop	Examples (references)
Vertebrates	Trp	Met	–	Stop	–	–	Human (1), bovine (2), rat (3), mouse (4), chicken (5), toad (6)
Arthropods	Trp	Met	–	Ser	–	–	*Drosophila* spp. (7, 8), mosquito (9, 10), 13 insect spp. (11)
Tunicates	Trp	Met	–	Gly	–	–	*Halocynthia roretzi* (12)
Echinoderms	Trp	–	Asn	Ser	–	–	Sea urchin (13, 14), Starfish (15, 16, 17, 18)
Molluscs	Trp	Met	–	Ser	–	–	Squid (19, 20), *Mytilus edulis* (21)
Nematodes	Trp	Met	*	Ser	*	*	*Ascaris suum* (22–24), *Caenorhabditis elegans* (22–24)
Platyhelminths	Trp	–	Asn	Ser	–	Tyr?	*Fasciola hepatica* (24a), planaria (24b)
Coelenterates	Trp	*	*	–	*	*	*Hydra* (22, 24), *Metridium senile* (22, 24)
Yeasts	Trp	Met	–	–	Thr	–	*Saccharomyces cerevisiae* (25, 26, 27, 28), *Torulopsis glabrata* (29)
Euascomycetes	Trp	–	–	–	–	–	*Aspergillus nidulans* (30, 31, 32), *Neurospora crassa* (33, 34)
Protozoans	Trp	–	–	–	–	–	*Trypanosoma brucei* (35), *Paramecium* spp. (36, 37, 38)

* Not determined; —, same as universal code. References cited: 1, Anderson *et al.* (1981); 2, Anderson *et al.* (1982); 3, Gadaleta *et al.* (1989); 4, Bibb *et al.* (1981); 5, Desjardins and Morais (1990); 6, Roe *et al.* (1985); 7, Clary and Wolstenholme (1985); 8, Garesse (1988); 9, HsuChen and Dubin (1984); 10, HsuChen *et al.* (1983); 11, Liu and Beckenbach (1992); 12, Yokobori *et al.* (1993); 13, Cantatore *et al.* (1989); 14, Jacobs *et al.* (1988); 15, Araki *et al.* (1988); 16, Himeno *et al.* (1987); 17, Jacobs *et al.* (1989); 18, Smith *et al.* (1990); 19, Watanabe, K. *et al.* (unpublished); 20, Shimayama *et al.* (1990); 21, Hoffmann *et al.* (1992); 22, Wolstenholme (1992); 23, Wolstenholme *et al.* (1987); 24, Wolstenholme *et al.* (1990); 24a, Garey and Wolstenholme (1989); 24b, Bessho *et al.* (1991); 25, Hensgens *et al.* (1979); 26, Hudspeth *et al.* (1982); 27, Macino *et al.* (1979); 28, Macino and Tzagoloff (1979); 29, Ainley *et al.* (1985); 30, Grisi *et al.* (1982); 31, Netzker *et al.* (1982); 32, Waring *et al.* (1981); 33, Bonitz and Tzagoloff (1980); 34, Browing and RajBhandary (1982); 35, Hensgens *et al.* (1984); 36, Pritchard *et al.* (1990*a*); 37, Pritchard *et al.* (1990*b*); 38, Seilhamer and Cummings (1982).

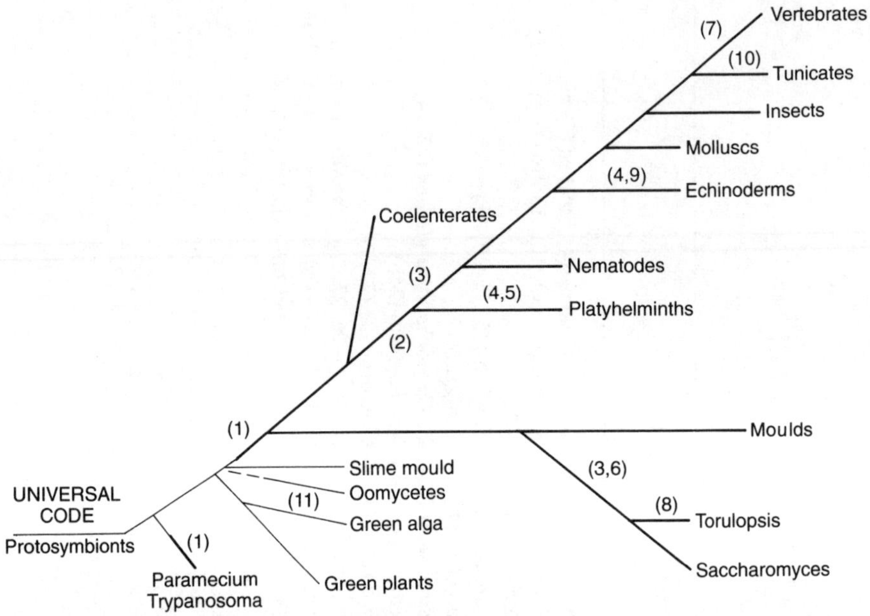

Fig. 6.14. Evolution of the genetic code in mitochondria. (1) UGA, Trp; (2) AGR, Ser; (3) AUA, Met; (4) AAA, Asp; (5) UAA, Tyr ?; (6) CUN, Thr; (7) AGR, stop; (8) CGN, non-coding; (9) AUA, to Ile from Met; (10) AGR, Gly. (11) CGG, UGA, UAG, non-coding in the green alga *Prototheca wickerhamii*. The point of change (3) is not definite. The branch lengths are diagrammatic and not to scale (from Jukes and Osawa (1993) with modifications; early branching orders are after Hasegawa *et al*., (1993) and T. Miyata (personal communication)).
——— universal genetic code; ——— non-universal genetic code.

The tRNAs of plant mitochondria, in which the universal code is used, have the usual cloverleaf secondary and L-shaped tertiary structures in common with bacterial and eukaryotic cytosolic tRNAs. Some of the tRNAs are the descendants of those in the original symbiotic eubacteria, which are ancestors of the present mitochondria. Some others have features in common with chloroplast tRNAs, and still others are of nuclear origin (Maréchal *et al.*, 1985; Jukes and Osawa, 1990; Maréchal-Drouard *et al.*, 1990).

Table 6.2 and Figs 6.1 and 6.14 summarize the reported non-universal genetic code in mitochondria. In most cases, the structure of the corresponding tRNA (or its gene) is known, so that it is possible to discuss the codon–anticodon interactions in some cases. In other cases, the information necessary to deduce the 'non-standard' codon–anticodon interactions of the modified nucleosides in anticodon is lacking.

UGA for Trp In all non-plant mitochondria except oömycetous mitochondria (Karlovsky and Fartmann, 1992) and slime mould mitochondria (Angata *et al.*,

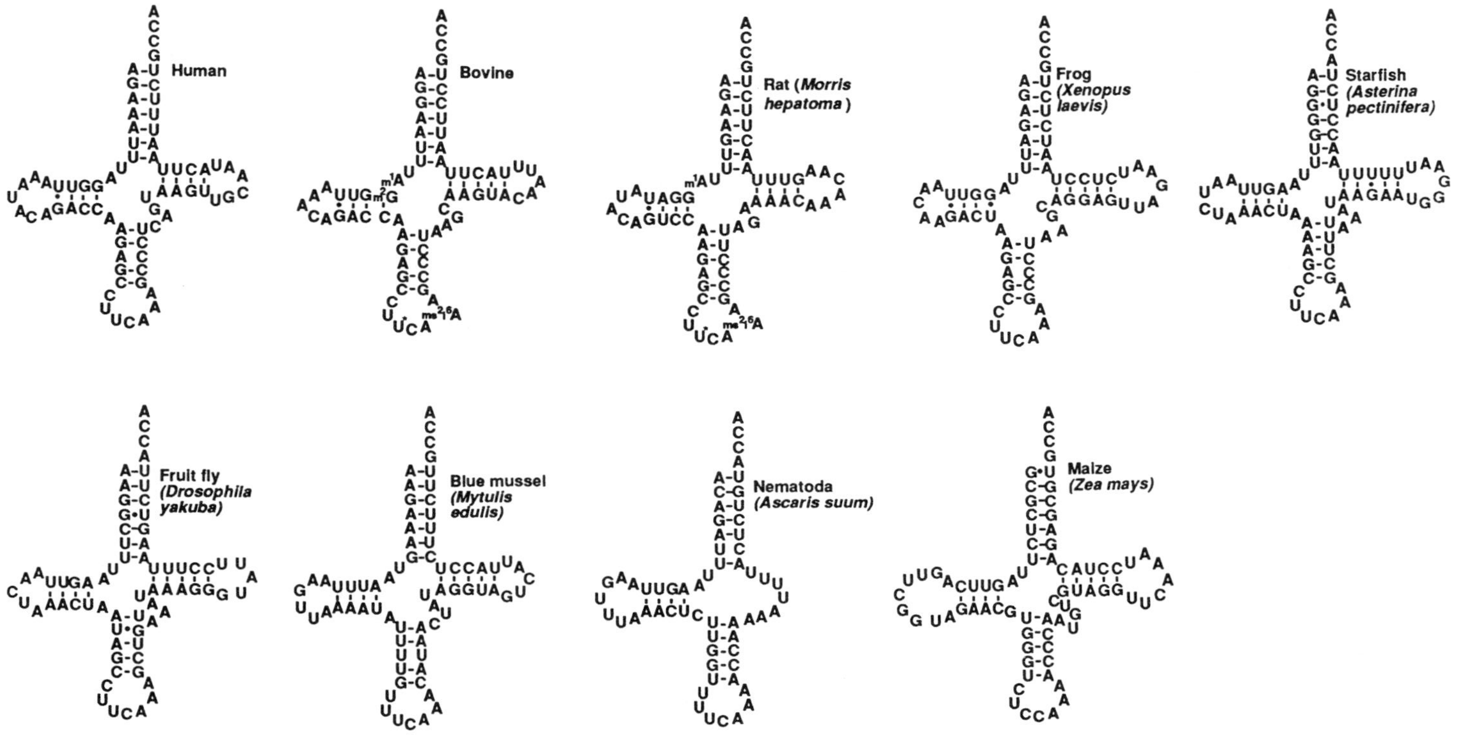

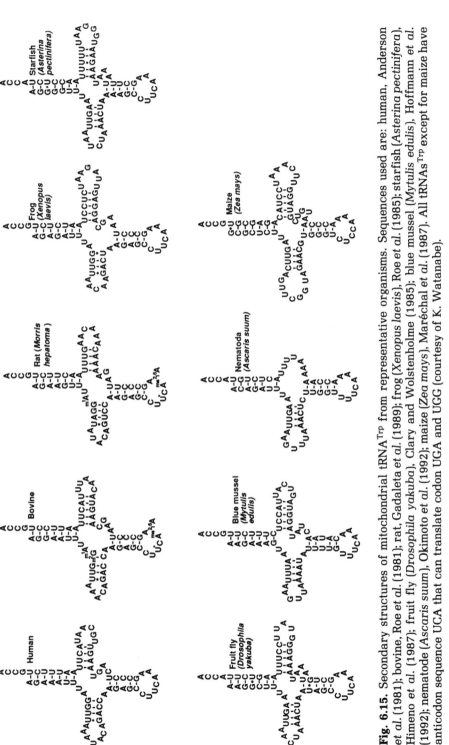

Fig. 6.15. Secondary structures of mitochondrial $tRNA^{Trp}$ from representative organisms. Sequences used are: human, Anderson *et al.* (1981); bovine, Roe *et al.* (1981); rat, Gadaleta *et al.* (1989); frog (*Xenopus laevis*), Roe *et al.* (1985); starfish (*Asterina pectinifera*), Himeno *et al.* (1987); fruit fly (*Drosophila yakuba*), Clary and Wolstenholme (1985); blue mussel (*Mytulis edulis*), Hoffmann *et al.* (1992); nematode (*Ascaris suum*), Okimoto *et al.* (1992); maize (*Zea mays*), Maréchal *et al.* (1987). All $tRNAs^{Trp}$ except for maize have anticodon sequence UCA that can translate codon UGA and UGG (courtesy of K. Watanabe).

1995), UGA is not a termination codon but is used as a Trp codon. In these mitochondria, $tRNA^{Trp}$ with the anticodon sequence of UCA or NCA (N: modified U), or the gene for $tRNA^{Trp}$ with the anticodon TCA, translate codons UGA and UGG by wobbling (Sprinzl *et al.*, 1989) (Fig. 6.15). Most of the Trp sites in the genes of *Schizosaccharomyces pombe* mitochondria are coded for by TGG, and only three are coded for by TGA in intronic open reading frames. The $tRNA^{Trp}$ has anticodon CCA (not UCA), which translates the UGA codon to Trp *in vitro* at a low frequency (Dirheimer and Martin, 1990). Note that some CNN anticodons translate codon NNA in addition to NNG when C is modified (see below).

In the oömyceteous mitochondrial COX3 gene, UGG is used exclusively as a Trp codon, even under very high AT-pressure (Karlovsky and Fartmann, 1992). Codon UGG is also used for Trp, and $tRNA^{Trp}_{CCA}$ has been detected, in slime mould mitochondria. Neither the Trp codon UGA nor $tRNA^{Trp}_{UCA}$ seems to exist (Angata *et al.*, 1995).

AUA for Met AUA, a universal Ile codon, is used as Met in yeasts and metazoan mitochondria, except echinoderms (Himeno *et al.*, 1987; Jacobs *et al.*, 1988; Cantatore *et al.*, 1989) and platyhelminths (Bessho *et al.*, 1991), in which AUA is an Ile codon. The anticodon of all $tRNA^{Met}$ genes is CAU, which, in all nuclear systems, translates codon AUG as Met (Sprinzl *et al.*, 1989).

Why can the $tRNAs^{Met}_{CAU}$ in most mitochondria translate both AUG and AUA codons as Met, whereas those from echinoderms and platyhelminths pair only with codon AUG? HsuChen *et al.* (1983) suggested that C in the wobble position, when unmodified, functions like modified U, and so can wobble-pair with A and G. However, such a wobble-pairing has not been reported. Sibler *et al.* (1985) speculated that an extra unpaired nucleotide within the base-paired T-stem may inlfuence the pairing properties of the CAU anticodon of yeast $tRNA^{Met\text{-}m}$ by conferring the ability to translate codons AUG and AUA. Similar unpaired nucleotides in the T-stem are found in the $tRNAs^{Met}$ of many animal mitochondrial genes.

The blue mussel, *Mytilus edulis*, possesses two $tRNA^{Met}$ genes with the anticodon sequences CAT and TAT (Hoffmann *et al.*, 1992). Perhaps AUR is translated by $tRNA^{Met}_{UAU}$, and AUG by $tRNA^{Met}_{CAU}$. However, $tRNA^{Met}_{UAU}$ has not been found in other mitochondria. A novel nucleoside—5-formylcytidine (f^5C)—has been found at the first anticodon position in mitochondrial bovine $tRNA^{Met}$, and in $tRNA^{Met}$ from the nematode *Ascaris suum*. This modification would made it possible for C to wobble-pair with AUG and AUA (Moriya *et al.*, 1994).

AAA for Asn In echinoderms (Himeno *et al.*, 1987; Jacobs *et al.*, 1988; Cantatore *et al.*, 1989) and platyhelminths (Ohama *et al.*, 1990*b*; Bessho *et al.*, 1991), AAA is a codon for Asn, and not for Lys (Table 6.3). All $tRNAs^{Asn}$ have the anticodon GTT at the DNA level, and these $tRNAs^{Asn}$ usually translate codons AAC and AAU by wobbling, following the standard wobble rules. However, the

Table 6.3 Comparison of AAA and AAG sites in flatworm and echinoderm mitochondria (mt) genes with corresponding sites in other mitochondria

	Genes and positions					
	COI gene			NDI gene		
Organisms	10,317	160,211	35,53,67	44	109	135
Flatworm	AAG	AAA	AAG	AAG	AAA	AAA
Sea urchin	AAG	AAA	AAG	AAA	Gln	AAA
Starfish			AAG	AAG	AAA	Asn
Yeast	Lys	Asn				
Trypanosome	Lys	Asn				
Six metazoans [1]	Lys	Asn	Lys	Lys	Asn	Asn

[1] *Drosophila, Xenopus*, mouse, rat, bovine, human.
From Jukes and Osawa (1991).

$tRNA^{Asn}$ of echinoderms and platyhelminths should also translate AAA as Asn. Thus, in these cases the wobble base may have been modified to I or equivalent, so as to pair with U, C, and A at the first codon position. It is significant that most mitochondrial $tRNA^{Lys}$ for both AAG and AAA codons have the anticodon UUU, whereas those for the AAG codon only, in echinoderms (Himeno *et al.*, 1987; Jacobs *et al.*, 1988) and *Fasciola* (a platyhelminth) (Garey and Wolstenholme, 1989) have the anticodon CUU pairing only with AAG. As an exception, mitochondrial $tRNA^{Lys}$ from the fruit fly has the anticodon CTT, which is thought to translate both AAG and AAA codons (HsuChen *et al.*, 1983). Its first anticodon nucleoside C might be modified to f^5C so as to pair with both G and A as in the case of $tRNA^{Met}_{CAU}$ of bovine and *Ascaris* mitochondria.

AGR for Ser, Gly, or stop The assignment of the AGA codon as Ser was first proposed by de Bruijn (1983) for mitochondria from *Drosophila melanogaster*, and then *D. yakuba* (Clary and Wolstenholme, 1985), the mosquito *Aedes albopiktus* (Dubin and HsuChen, 1984), starfish (Himeno *et al.*, 1987), and sea urchins (Jacobs *et al.*, 1988; Cantatore *et al.*, 1989). Echinoderm and some insect mitochondria have been known to use AGG as Ser, in addition to AGA. It is now evident that codons AGA and AGG (AGR) are used as Ser in most invertebrate mitochondria, except for cnidaria, in which AGR is read as Arg (Wolstenholme *et al.*, 1990; Wolstenholme, 1992). In all metazoans except cnidaria, $tRNA^{Ser}_{GCU}$ is thought to be responsible for translation of all four AGN codons as Ser, as no other $tRNA^{Ser}$ exists. Only in *Ascaris suum* mitochondria does $tRNA^{Ser}_{UCU}$ which seems to translate all AGN codons as Ser by 4-way wobbling (Wolstenholme *et al.*, 1987; Okimoto *et al.*, 1992).

Figure 6.16 shows the secondary structures of $tRNA^{Ser}_{GCU}$ in mitochondria from various organisms. As the G at the first anticodon position cannot usually pair

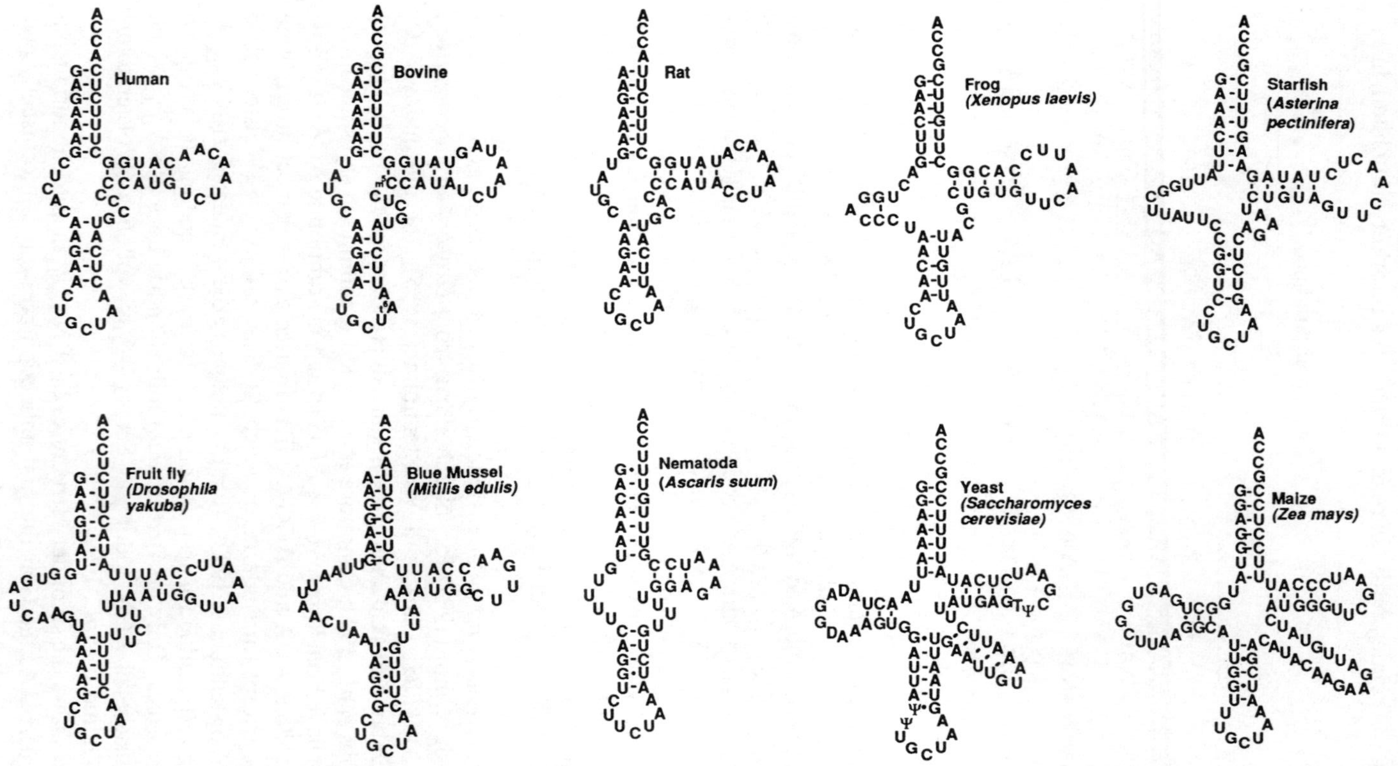

Fig. 6.16. Secondary structures of mitochondrial tRNA^{Ser} from representative organisms. Sequences used are: human, Anderson *et al.* (1981); bovine, Roe *et al.* (1981); rat, Gadaleta *et al.* (1989); frog (*Xenopus laevis*), Roe *et al.* (1985); starfish (*Asterina pectinifera*), Himeno *et al.* (1987); fruit fly (*Drosophila yakuba*), Clary and Wolstenholme (1985); blue mussel (*Mytulis edulis*), Hoffmann *et al.* (1992); nematode (*Ascaris suum*), Okimoto *et al.* (1992); yeast (*Saccharomyces cerevisiae*), Martin *et al.* (1982); maize (*Zea mays*), Wintz *et al.* (1988) (courtesy of K. Watanabe).

with the A or G in the codon wobble positions, G at the first anticodon position might be modified so as to pair with all AGN codons. If this is not the case, a region, or regions, other than the anticodon would influence the decoding ability of the AGR codons. The first likely candidate for this is the D-arm, which varies in size according to the evolutionary position of the animal in question. The $tRNA^{Ser}_{GCU}$ of vertebrate mitochondria lacks almost the entire D-arm (5 to 10 residues in this region), whereas the invertebrates have D-arms of intermediate sizes (more than 11 residues) with, however, the exception of *Fasciola* mitochondrial $tRNA^{Ser}_{GCU}$ (which has a D-arm with seven residues, which is still thought to decode AGR codons) (Garey and Wolstenholme, 1989), and the fish *Cyprinus carpio* $tRNA^{Ser}_{GCU}$ (which has an intermediate-sized D-arm (13 residues) (Chang, Y. S. and Huang, F. L., EMBL accession no. ×61010, unpublished), and yet apparently decodes only AGY codons).

The second likely candidate is the G–C pair at the bottom of the anticodon stem, which is present in the $tRNA^{Ser}_{GCU}$ of invertebrate mitochondria, but not in vertebrate mitochondria, in which its place is taken by the A–U pair (Osawa *et al.*, 1992), except for hamster mitochondrial $tRNA^{Ser}_{GCU}$ (Baer and Dubin, 1980).

In the cytochrome oxidase subunit I (COI) gene of ascidian mitochondria (*Halocynthia roretzi* and *Pyura stolonifera*), codons AGA and AGG quite often appear in the reading frame. Sequence comparisons with the corresponding regions of other animal mitochondrial COI genes indicate that the AGA and AGG codon sites in the ascidian mitochondrial genome are mainly occupied by GGN Gly, and not by Ser codons (in most invertebrates) or stop codons (in vertebrates). This suggests that, in addition to GGN, AGA and AGG are Gly codons in the ascidian mitochondria (Yokobori *et al.*, 1993; Watanabe, K., personal communication). In support of this view, the gene for $tRNA^{Gly}$ with anticodon UCU, which would read AGR codons, has been identified (Watanabe, K., personal communication).

AGA and AGG in vertebrate mitochondria are used as stop codons; neither of those codons occurs internally in any of the protein genes, and tRNAs translating these codons do not exist (Anderson *et al.*, 1981, 1982; Bibb *et al.*, 1981; Roe *et al.*, 1985; Gadaleta *et al.*, 1989). Release factor of rat mitochondria recognizes only UAR stop codons and not AGR *in vitro* (Lee *et al.*, 1987). Further studies are needed to clarify this.

CUN for Thr All CUN codons in *Saccharomyces cerevisiae* mitochondria are translated as Thr instead of Leu. Li and Tzagoloff (1979) identified the relevant $tRNA^{Thr}$ gene with the anticodon sequence of TAG in the 8-base anticodon loop. Later, it was pointed out that, if a U : U pair is added to the anticodon stem and a 6-base anticodon loop is permitted, *S. cerevisiae* $tRNA^{Thr}_{UAG}$ would have an anticodon loop structure similar to that of *Torulopsis glabrata* $tRNA^{Thr}_{UAG}$, which has a normal 7-base anticodon loop and translates CUN codons as Thr (Osawa *et al.*, 1990*a*) (Fig. 6.17). With such an anticodon loop structure, the yeast $tRNA^{Thr}_{UGA}$ would base-pair with any of four CUN codons by 4-way wobbling.

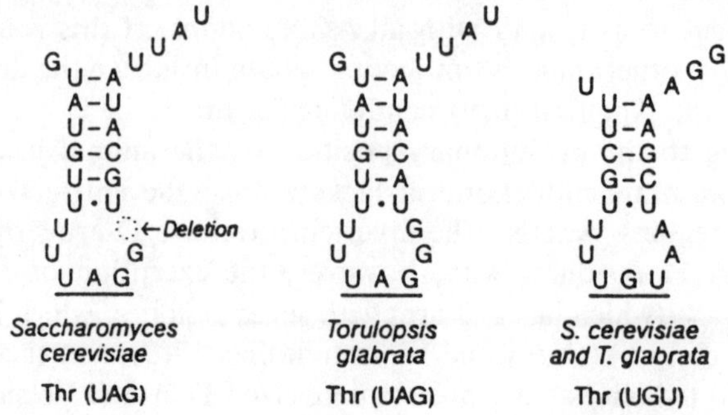

Fig. 6.17. Similarity in the anticodon arm structures of $\mathrm{tRNA}^{\mathrm{Thr}}_{\mathrm{UAG}}$ from two yeast species, *Saccharomyces cerevisiae* and *Torulopsis glabrata*. The same region for $\mathrm{tRNA}^{\mathrm{Thr}}_{\mathrm{UGU}}$ is shown for comparisons (from Osawa *et al.*, 1990*a*).

Two different threonyl-tRNA synthetases were found in the yeast nuclear genome; one is for the above $\mathrm{tRNA}^{\mathrm{Thr}}_{\mathrm{UAG}}$ and the other for $\mathrm{tRNA}^{\mathrm{Thr}}_{\mathrm{UGU}}$ corresponding to ACN, the universal Thr codons. One of them—MST-1—aminoacylating only $\mathrm{tRNA}^{\mathrm{Thr}}_{\mathrm{UAG}}$, has a high sequence homology to another threonyl-tRNA synthetase for $\mathrm{tRNA}^{\mathrm{Thr}}_{\mathrm{UGU}}$ and also to those from bacteria, so that it would seem to have resulted from duplication of the gene for the original threonyl-tRNA synthetase for $\mathrm{tRNA}^{\mathrm{Thr}}_{\mathrm{UGU}}$ (Pape and Tzagoloff, 1985). The new threonyl-tRNA synthetase would have acquired the recognition ability towards $\mathrm{tRNA}^{\mathrm{Thr}}_{\mathrm{UAG}}$ by changing its structures (Osawa *et al.*, 1992).

UAA for Tyr UAA may be used as Tyr in the planarian *Dugesia japonica*, because COI gene contains UAA at the well-conserved Tyr site of UAY (Bessho *et al.*, 1991). No characterization for its corresponding $\mathrm{tRNA}^{\mathrm{Tyr}}_{\mathrm{UUA}}$ was reported.

The changes in the mitochondrial genetic code may be summarized in the form of a phylogenetic tree (see Fig. 6.14). This scheme assumes that the endosymbiotic acquisition of mitochondria by eukaryotes occurred only once.

The mitochondria of green plants, oömycetes, and slime moulds use the universal genetic code. The first change—codon UGA from stop to Trp—occurred in some protozoans. The same change took place independently in non-plant mitochondria after the separation from plant/oömycetes/slime mould mitochondria. *Paramecium/Trypanosoma* branched off after fungi in a 5S rRNA tree (Hori and Osawa, 1987). In that case, the change of UGA from stop to Trp occurred once. An alternative explanation would be that the mitochondrial ancestor used UGA as a Trp codon and in the ancestor leading to green plants and a few others, UGA was reassigned to stop (Boyen *et al.*, 1994). In mitochondria of the red alga *Chondrus crispus*, UGA is a Trp codon (Boyen *et al.*, 1994), which is not shown in this figure because the phylogenetic position of the alga is not definite. The green plant mitochondrial code would probably be an evolutionary mosaic, as seen in the composition of tRNA (see p. 86) and others. Thus, it may not be appropriate to place the green plant code in a single phylogenetic tree. In the animal line, AGR Arg codons became Ser, starting with the platyhelminths and continuing in nematodes, echinoderms, molluscs, and insects, until AGR became stop in vertebrates. However, AGR changed uniquely from Ser to Gly in the ascidian branch. AAA changed from Lys to Asn in the branch leading to platyhelminths and, as a separate event, in echinoderms.

UAA stop probably codes for Tyr in platyhelminths. Starting with nematodes, AUA changed from Ile to Met. This change was reversed independently in platyhelminths and in echinoderms.

In the fungal line, changes in the yeast mitochondrial code are AUA from Ile to Met (which occurred independently from that in the animal mitochondrial code) and CUN from Leu to Thr. No changes from the universal code, other than UGA to Trp, have taken place in euascomycetes.

6.4 Mechanisms of code change

6.4.1 *Codon capture theory*

If the change in meaning of a codon, e.g. AAA Lys to Asn, were to occur directly, the tRNA that translated AAA to Lys would have to change its identity determinants to react with aminoacyl-tRNA synthetase for Asn, and Asn would have to be inserted instead of Lys at all sites previously occupied by AAA for Lys. As a result, the nucleotide sequences would not change, but the amino acid sequences in proteins would. Such direct replacements of one amino acid by another would be disruptive, as Crick discussed when he proposed his frozen-accident theory. Even in mitochondria, where only a small number of proteins are synthesized, such replacements could be lethal.

A more plausible possibility is that the codon reassignment follows the codon capture rule (Osawa and Jukes, 1989), in which a series of non-disruptive, i.e. neutral, changes is involved. This theory proposes the temporary disappearance of an amino acid codon (or stop codon) from coding frames by conversion to another synonymous codon, and a loss of the corresponding tRNA that translates the codon. For a stop codon, the release factor must change simultaneously so as not to recognize the stop codon. This change will produce an unassigned codon. Therefore, the production of unassigned codons (discussed in Sections 5.1 and 5.2) is probably an intermediate step in codon reassignment during evolution of the genetic code. As discussed above, unassigned codons may be produced by directional mutation pressure, changes of release factor (RF) or genomic economization. The codon reappears later by conversion of another codon (mainly a codon that is synonymous with the altered codon) and emergence of a tRNA (or RF) that translates (or recognizes) the reappeared codon with a different assignment. As a result, the nucleotide sequences change but the amino acid sequences of proteins do not change. This view is supported by the fact that all the reported identifications of non-universal codons involve the DNA sequence comparisons of genes containing an expected non-universal codon with their counterparts in other organisms (or mitochondria) or with one of the duplicated genes in the same organisms (or mitochondria), in which the universal codon is used. The predicted non-universal codon, e.g. AAA for Asn (Lys in the universal genetic code) in certain mitochondria, occurs mostly at sites corresponding to the universal AAY Asn sites (but not to the universal AAR Lys sites) in other mitochondria (see Table 6.3). Naturally the DNA sequence comparisons alone involve potential pitfalls, such as RNA editing (see Sections 8.2 to 8.6). The cDNA sequence comparisons are more desirable in this respect. At any rate, the above descriptions, together with the individual examples described below, indicate that no direct replacement of codon A for amino acid *a* by the altered codon A for amino acid *b* occurs during amino acid assignment change from *a* to *b* of the codon.

6.4.2 *Codon capture*

The unassigned codons could conceivably be reassigned to the same amino acid (or stop signal), or to another amino acid, depending on the appearance of a new tRNA (or RF).

The unassigned codon capture may be implemented in one of the following ways: (1) by a change in the anticodon; (2) by a change to a different amino acid in aminoacylation of a tRNA molecule, while retaining the same anticodon; (3) in mitochondria, by a change in codon–anticodon pairing (Osawa *et al.*, 1992; see also Jukes and Osawa, 1993; Ueda and Watanabe, 1993).

In 1963, before the universal genetic code was established, Crick had already noted:

It used to be argued that once a complete code had arisen it would be very difficult to change since any alteration in the meaning of a triplet would produce changes in almost every protein in the organism and would thus in all probability be lethal. This argument would lose some of its force if the code were degenerate and if one particular triplet were used very infrequently. For instance, in an organism having a DNA very high G and C the triplet AAA would tend to be rare. Under these circumstances mutations in the sRNA (=tRNA) and the appropriate activating enzyme, which would allow it to change its meaning, might not be lethal. If, in the course of further evolution, the base ratio of the DNA of such an organism drifted back from the extreme value to one having more A and T, then the triplet AAA might become fairly common . . .

Surprisingly, this argument is almost in the line of the codon capture theory: it clearly states that the codon reassignment may occur by directional mutation pressure (which is changeable) and by the procedures of (1) and/or (2) mentioned above. However, Crick was inclined to believe that the code did not change, because:

Fortunately, in the last year or so, experimental evidence has accumulated suggesting that the code may well be universal. The method used is to synthesize a particular protein in a

Table 6.4 Possible changes of AUN codons during evolution

Anticodon	Codon	Amino acid	Remarks
GAU	AU$^{U}_{C}$	Ile	
(AAU	AUU	Ile)	
UAU	AU$^{A}_{G}$	Met	
(CAU	AUG	Met)	
	Changing to		
(GAU	AU$^{U}_{C}$	Ile)	
((AAU	AUU	Ile))	
IAU	U AUC A	Ile	i.e. AUA codes both Ile *and* Met
UAU	AU$^{A}_{G}$	Met	
CAU	AUG	Met	
	Changing to		
(GAU	AU$^{U}_{C}$	Ile)	
((AAU	AUU	Ile))	No ambiguity; the only codon for Met is now AUG
IAU	U AUC A	Ile	
CAU	AUG	Met	

Parentheses show optional cases.
From Crick (1965), cited in Osawa *et al.* (1989*b*).

cell-free system, using some of the components from one species and some from another (including *E. coli*, *Micrococcus* (!), yeast, TMV, or phage f2, etc.) . . . All these experiments suggest that the code may well be universal, but they might also be compatible with codes that, while mainly universal, differ slightly from one organism to another.

In 1965, when the universal genetic code was almost established, Crick proposed the frozen-accident theory. At that time Crick still seemed to suspect that the code change might occur and he noted that the codes for AUN might change during evolution as shown in Table 6.4 (Crick, 1965).

The first change could come about because the A in the anticodon [AAU] was deaminated to I. This would make AUA ambiguous but it would not necessarily be lethal. The triplet AUA would tend to be selected against, and thus drop out of use, especially in an organism with high (G + C) in its DNA. Note that at this stage both Ile and Met have at least one *un*ambiguous codon. Finally the anticodon UAU for Met might be eliminated, and the codon AUA taken over by Ile unambiguously.

This illustration shows that for certain triplets the code may not be as difficult to change as we thought. Note that the change can also proceed in the reverse direction (Crick, 1965).

This is the same line of discussion as in 1963. In this proposal, which was omitted from his subsequent publication in 1968, Crick clearly stated the idea that GC-pressure might lead to the elimination of the ambiguous codon AUA, and that this codon might reappear with a different assignment, because 'the change can also proceed in the reverse direction'.

Throughout, Crick supposed that some ambiguity would be allowed transitionally during the course of codon reassignment. For example, in Table 6.4 GC-pressure makes AUA a rare codon, which could be recognized by two tRNAs, mainly by $tRNA^{Ile}_{IAU}$ and partly by $tRNA^{Met}_{UAU}$.

In contrast, the codon capture theory assumes the complete disappearance of a codon together with removal of the corresponding tRNA, producing an unassigned codon. Hence there would be no ambiguity. In some code changes, some ambiguity might be acceptable, as argued by Crick and, more recently, by Schultz and Yarus (1994), although there are no such examples in natural populations.

6.5 Nuclear systems

6.5.1 *UGA from stop to Trp*

The appearance of the UGA Trp codon in the *Mycoplasma* lineage would have taken place via the following steps (Fig. 6.18) (Jukes, 1985; Osawa *et al.*, 1990*a*,*b*, 1992). AT-pressure led to the replacement of all UGA stop codons by UAA. In A+T-rich bacteria, the use of UGA as stop codons is rare, and most stop codons are UAA even in *E. coli* (G+C content = 50 per cent). Therefore, complete conversion of UGA to UAA could take place. At about this stage, RF-2, which interacts with UAA and UGA, would have either been deleted or become specific for UAA, so that UGA would have become an unassigned codon. Then $tRNA^{Trp}_{CCA}$ duplicated, and the anticodon of one of the duplicates, under AT-pressure, mutated to UCA. The tandem arrangement of the genes for $tRNA^{Trp}_{UCA}$ and $tRNA^{Trp}_{CCA}$ on the chromosome of this species strongly supports this (Yamao *et al.*, 1988) (Fig. 6.19). This change would not have taken place in such a bacterium as *Acholeplasma laidlawii* (carrying only $tRNA^{Trp}_{CCA}$) (Tanaka *et al.*, 1989), which shares a common ancestor with *Mycoplasma* spp. The new Trp anticodon could pair with both UGA and UGG, so that UGA could be produced by the AT-pressure that caused mutation of some Trp codons from UGG to UGA. Note that, before the emergence of $tRNA^{Trp}_{UCA}$, the mutations of UGG to UGA, even under AT-pressure, could not have taken place. Such mutations would have been removed by negative selection because of the lack of tRNA translating UGA codons.

As anticodon UCA pairs with both UGA and UGG, anticodon CCA is no longer needed, although it is still present in *Mycoplasma capricolum*. However, $tRNA^{Trp}_{CCA}$ is charged much less by Trp in the cells than $tRNA^{Trp}_{UCA}$, and the intercellular amount of $tRNA^{Trp}_{CCA}$ is 5 to 10 times lower than that of $tRNA^{Trp}_{UCA}$. The predominance of $tRNA^{Trp}_{UCA}$ over $tRNA^{Trp}_{CCA}$ results from the strong attenuation of transcription by the terminator-like structure present between the two genes

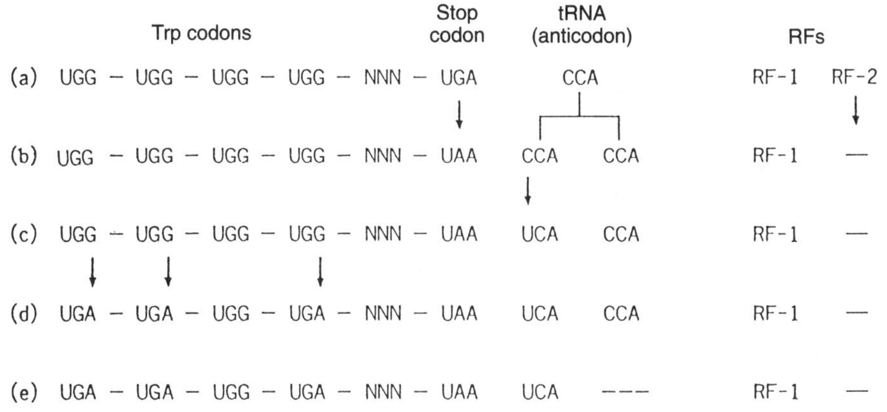

Fig. 6.18. A model for evolution of the UGA Trp codon in Mollecutes. Stages (a), (d), and (e) are represented by *Acholeplasma laidlawii*, *Mycoplasma capricolum*, and *Mycoplasma pneumoniae/genitalium*, respectively (from Osawa *et al.*, 1992).

(Yamao *et al.*, 1988). It is possible that such a restriction will lead eventually to the disappearance of the $tRNA^{Trp}_{CCA}$ gene. In fact, the gene for $tRNA^{Trp}_{CCA}$ is not present in some of the *Mycoplasma* species, such as *Mycoplasma pneumoniae* and *Mycoplasma genitalium* (Inamine *et al.*, 1990). In this way, UGA would have been captured (reassigned) by Trp in *Mycoplasma* spp.

If the RF activity for UGA remains intact in the *Mycoplasma* lineage, UGA codons in the reading frames would be recognized by both the RF and $tRNA^{Trp}_{UCA}$. This would be disadvantageous because it would result in production of a mixture of truncated and complete peptides. In other words, the tRNA would be nothing but a suppressor. Inagaki *et al.* (1993) have constructed a cell-free translation system of *M. capricolum* using synthetic mRNA including codon UAA [mRNA (UAA)], UAG [mRNA (UAG)], and UGA [mRNA (UGA)] in-frame. In the absence of Trp, the translation of mRNA (UGA) ceased at the UGA sites without appreciable release of the synthesized peptides from the ribosomes, whereas with mRNA (UAA) or mRNA (UAG), the bulk of the peptides was released. Upon addition of the *E. coli* S-100 fraction or *B. subtilis* S-100 fraction to the translation system, the synthesized peptides with mRNA (UGA) were released almost completely from the ribosomes, presumably because of the presence of RF-2 active to UGA in the added S-100 fraction. These data suggest that RF-2 is deleted in *M. capricolum*, in accordance with the assumption that UGA had been unassigned before it became a Trp codon (Inagaki *et al.*, 1993) (Fig. 6.20).

A similar capture of UGA codon would have occurred in the ancestor of mitochondria. Rat mitochondria, in which UGA is a codon for Trp, lacks RF corresponding to the *E. coli* RF-2 (Lee *et al.*, 1987).

This type of codon reassignment is known as 'stop codon capture', because the former stop codon (in this case UGA) has been captured by an amino acid (Trp).

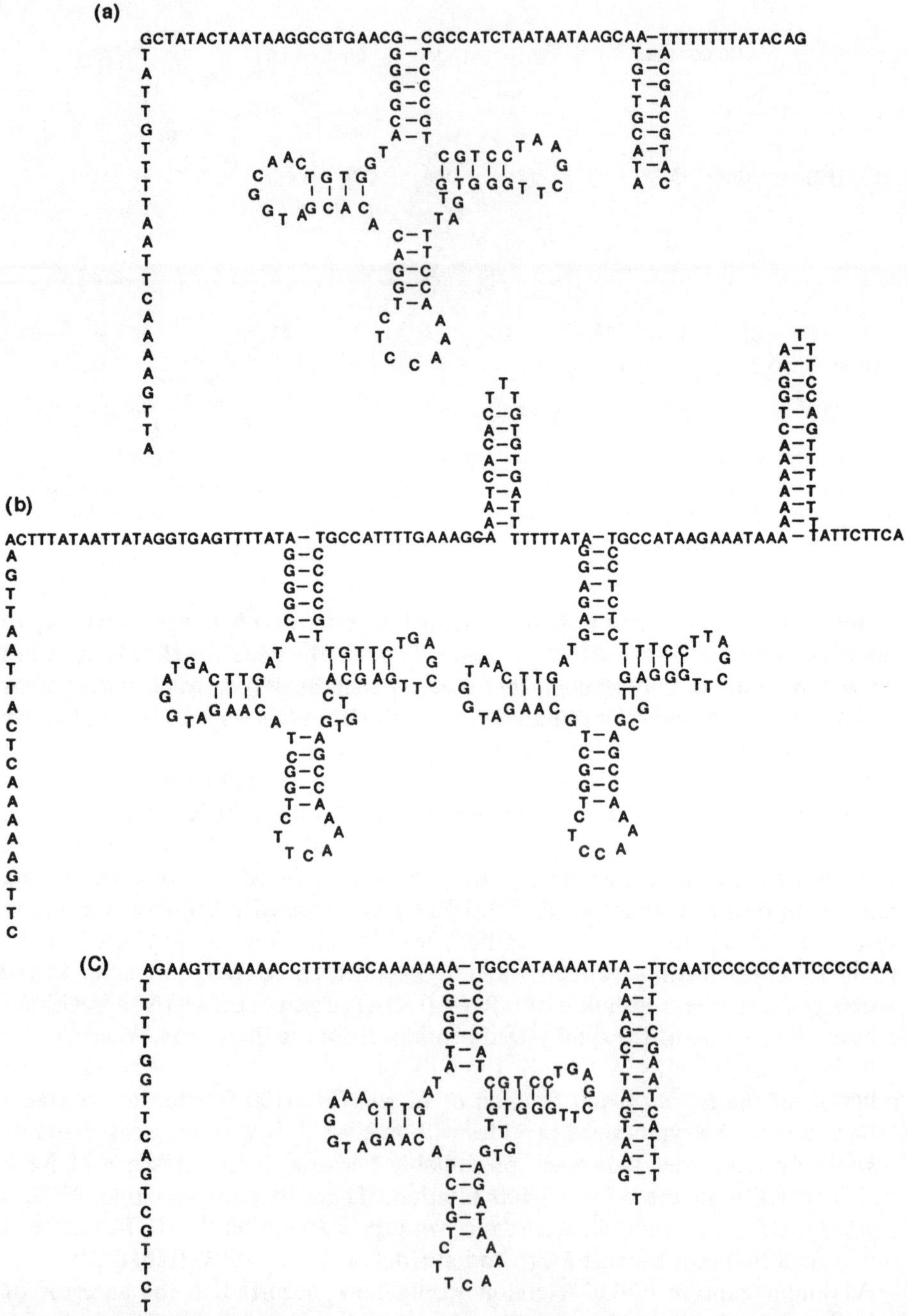

Fig. 6.19. tRNA^{Trp} genes from three species of Mollecutes. (a) *Acholeplasma laidlawii*; (b) *Mycoplasma capricolum*; (c) *Mycoplasma genitalium*. Note the presence of terminator-like structure between two genes in *Mycoplasma capricolum* (from Osawa *et al.*, 1990*b*).

A new tRNA (tRNA$^{Trp}_{UCA}$) is not a suppressor, as UGA is a regular Trp codon and is not used as a stop codon in this bacterium. Changes of UAR from stop to Gln, and of UGA from stop to Cys (to be discussed below) are also examples of stop codon capture.

6.5.2 *UAR from stop to Gln*

The UAR Gln codons in some ciliated protozoans (A + T content = 70 to 75 per cent), such as *Tetrahymena* and *Paramecium* spp., would have evolved as follows (Jukes *et al.*, 1987; Osawa and Jukes, 1988; Osawa *et al.*, 1992). UGA is the sole stop codon in these ciliates. In eukaryotes, only one species of RF recognizes all three stop codons. In the *Tetrahymena* lineage, the RF would have evolved to be specific to UGA; and UAA and UAG would have been removed from the termination sites and become unassigned. This process would have occurred apparently independently of AT-pressure, which cannot convert UAA to UGA. Then the gene for tRNA^{Gln}, with the anticodon UmUG, would have duplicated, and one of the duplicates would have mutated to anticodon UmUA. (If more than one gene for tRNA$^{Gln}_{UmUG}$ existed, one of them could have mutated to tRNA$^{Gln}_{UmUA}$.) Later, the gene for tRNA with anticodon UmUA would have duplicated again and, in one of the duplicates, anticodon UmUA would have mutated to CUA, thus improving the fidelity of codon–anticodon pairing with UAG. Under AT-pressure, some of the Gln codons CAA and CAG would have mutated to UAA and UAG, which are read as Gln. A similar reassignment of UAR to Gln could have occurred independently in *Acetabularia* spp. (Schneider *et al.*, 1989) having A + T-rich genomes. In these cases, very high AT-pressure would not be required, because recognition of UAR by RF would have been lost as a result of accumulation of mutations on RF.

6.5.3 *UGA from stop to Cys*

The UGA Cys codon in *Euplotes octacarinatus* (Meyer *et al.*, 1991) would have been created as follows (Jukes *et al.*, 1991). UAA is the stop codon; UAG has not been detected. We suggest that, in the *Euplotes* lineage, RF would have evolved to be specific for UAA (and perhaps UAG?), so that UGA disappeared and became unassigned. When G in the Cys anticodon GCA was modified, perhaps to I or equivalent so as to pair with U, C, and A in the third codon position of UGU, UGC, and UGA, respectively, UGA reappeared as a Cys codon.

6.5.4 *CUG from Leu to Ser and as a reversal to Leu*

As described in Section 6.3.1, the assignment of codon CUG changed from Leu to Ser, and from Ser to Leu again as a reversal during evolution of yeasts. The codon capture theory assumes that the production of unassigned codon would be the first step for neutral genetic code change. This would mean that CUG became unassigned twice, followed by the reassignment of the codon. Has directional

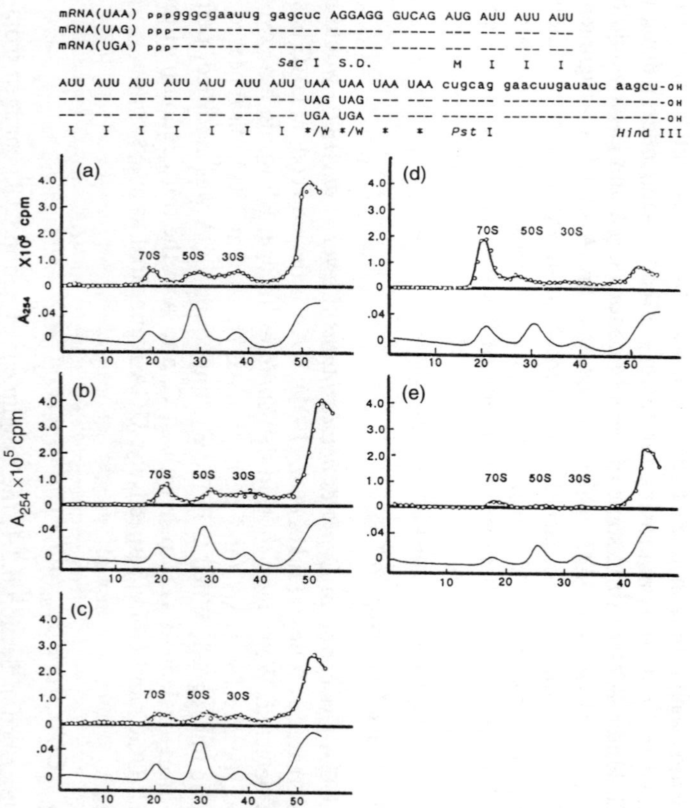
mRNA(UAA) pppgggcgaauug gagcuc AGGAGG GUCAG AUG AUU AUU AUU
mRNA(UAG) ppp
mRNA(UGA) ppp
Sac I S.D. M I I I
AUU AUU AUU AUU AUU AUU AUU UAA UAA UAA UAA cugcag gaacuugauauc aagcu-OH
UAG UAG
UGA UGA
I I I I I I I */W */W * * Pst I Hind III
(a)
(b)
(c)
(d)
(e)
70S
50S
30S
X10⁵ cpm
A₂₅₄
A₂₅₄ ×10⁵ cpm

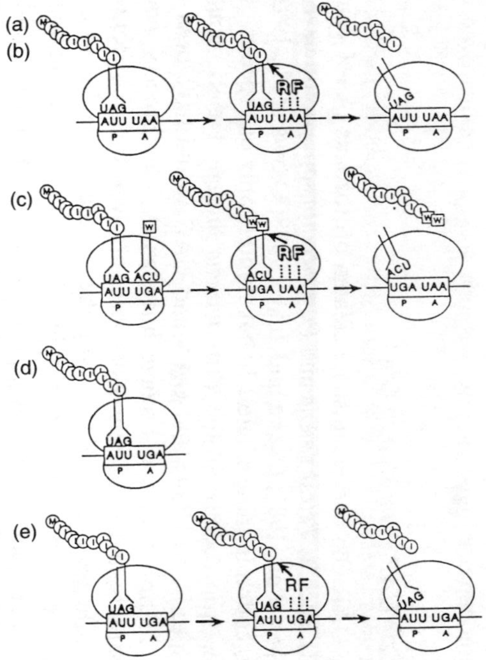
(a)
(b)
(c)
(d)
(e)
RF
UAG
ACU
AUU UAA
AUU UGA
UGA UAA
P
A
W

mutation pressure played a role in making yeast CUG an unassigned codon? A remarkable feature of the yeasts resides in their wide range of genomic G+C content (27.1 to 63.4 per cent), suggesting that the direction of the mutation pressure has changed rapidly within a very short time, even in closely related yeasts. It is thus possible that CUG became unassigned twice in two different periods under strong AT-pressure by converting to another A+T-rich synonymous codon, accompanied by a loss of the corresponding tRNA. A relaxation of AT-pressure, i.e. an increase in GC-pressure, could have triggered the reappearance of codon CUG upon emergence of a tRNA translating the codon.

The following is a plausible scenario for the codon changes (Fig. 6.21) (Ohama *et al.*, 1993; Watanabe *et al.*, 1993; unpublished data). First, codon CUG Leu disappeared, under strong AT-pressure, by conversion, e.g. to Leu UUR with removal of the $tRNA^{Leu}_{CAG}$ that translates codon CUG (Group I in Fig. 6.13). Then, with GC-pressure, the unassigned CUG codon so produced could have appeared as a Ser codon after the emergence of a new $tRNA^{Ser}$ with anticodon CAG for codon CUG (Group II). Naturally, such an unassigned codon capture by the $tRNA^{Ser}$ is acceptable at any dispensable site in a gene. For example, a dispensable UUG Leu could mutate to CUG Ser. However, CUG is used as a Ser codon in the catalytic centre of one of the lipase genes of *Candida cylindracea* (Kawaguchi *et al.*, 1989), showing that a Ser codon—even at an important site—can change to CUG by a nearly neutral process. The change of the universal Ser codon UCN or AGY to CUG cannot occur by a single mutation, and must pass through an intermediate codon such as UUR (Leu) or CCN (Pro). The gene with such an intermediate codon becomes functionally inferior or inactive if it occurs at the important site. However, most of the eukaryotic genes exist multiply in the genome, and therefore a deleterious mutation in one of the genes would not affect the viability of the cell. The 'inferior' gene would become a pseudogene at one time, but at another time it would be revived by a second mutation at the site originally occupied by a Ser codon, such as UCG [e.g. UCG (Ser)→UUG (Leu)→CUG (Ser)]. The latter process would not be impossible, if the creation of $tRNA^{Ser}_{CAG}$ preceded the change of UUG (Leu) to CUG (Ser). Indeed, there would

---

Fig. 6.20. Lack of recognition of codon UGA by RF-2 in *Mycoplasma capricolum*. Left: Sucrose-gradient centrifugation of reaction mixture labelled with [^{3}H]Ile. (a) Translation of mRNA(UAA); (b) translation of mRNA(UAG); (c) translation of mRNA(UGA) in the presence of Trp; (d) the same as (c) but in the absence of Trp; (e) translation of mRNA(UGA) in the absence of Trp and the presence of the *E. coli* S100 fraction.

Right: Models for cell-free translation of synthetic mRNAs. Models (a–e) are constructed to explain the results of (a–e) in the left-hand figures. Figures begin with the tenth AUU Ile codon on the P site of the ribosome, so that the first test codon is on the A site. RF in (a) and (b) (presumably RF-1) responds only to UAA and UAG. In (d), in the absence of Trp, no further reaction occurs because of the absence of the tryptophanyl-tRNA and RF (presumably RF-2) for UGA. In (e) RF (presumably RF-2) in the *E. coli* S100 fraction responds to UGA. The mRNAs used are shown above the figures (from Inagaki *et al.*, 1993).

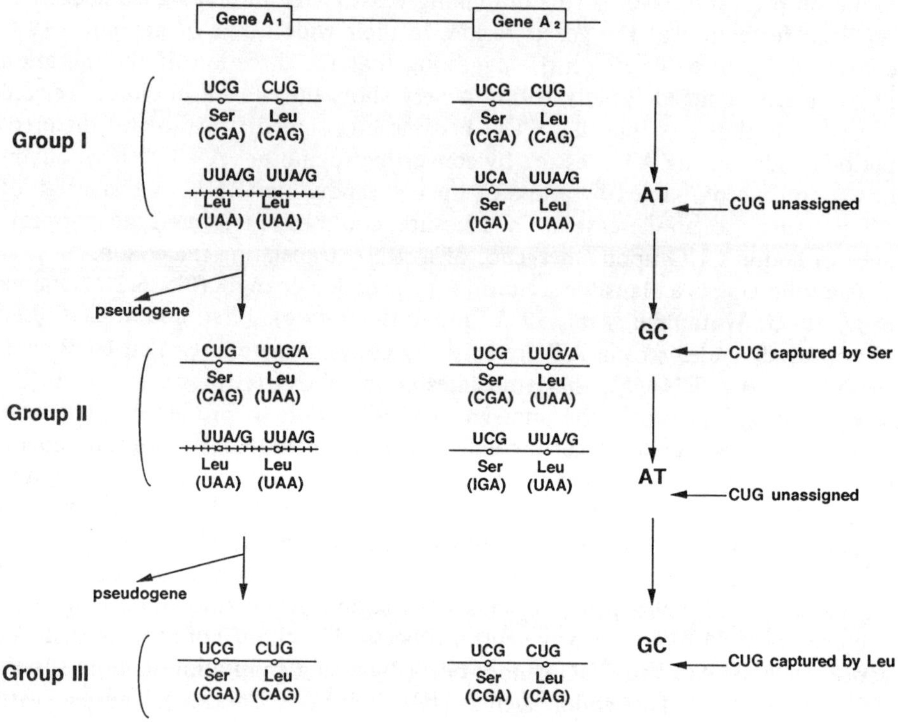

Fig. 6.21. A model for changes in the amino acid assignment of codon CUG during the evolution of yeasts. Only one each of the essential Ser and Leu sites in a gene is shown. Upper line, codon; lower line with parentheses, anticodon. active gene; inactive gene. NNA/G (or NNG/A) denotes that NNA (or NNG) predominates over NNG (or NNA) under AT-(or GC-) pressure. For further details see text.

be no other way of near-neutral generation of CUG Ser codons at the important sites.

Upon further evolution, the above process would have been reversed (Fig. 6.13; Group III). With strong AT-pressure, CUG Ser became unassigned by converting to an intermediate codon, such as UUG Leu, with the removal of $tRNA^{Ser}_{CAG}$. By relaxation of AT-pressure, i.e. the emergence of GC-pressure, some UUR Leu codons mutated to UCR Ser translatable with pre-existing $tRNA^{Ser}_{CGA}$ or $tRNA^{Ser}_{IGA}$, e.g. CUG (Ser)→UUG (Leu)→UCG (Ser). CUG would have appeared as a Leu codon from UUA or CUY/A Leu, without passing through an intermediate codon upon the emergence of $tRNA^{Leu}_{CAG}$ from a mutation of one of the $tRNAs^{Leu}_{IAG}$ (or $tRNAs^{Leu}_{UAG}$).

Naturally, the higher the GC-pressure, the higher the usage of codon CUG, regardless of its assignment—Ser or Leu. For example, CUG is a rare Leu codon in *S. cerevisiae* (G+C=40 per cent) and is also a rare codon (for Ser) in *C. albicans* (G+C=34 per cent), while it is a predominant Ser codon in *C. cylindracea* (G+C=63 per cent).

Propagation of codon CUG for Ser is also possible by duplication of the genes containing CUG after its reassignment.

It may be inferred that all of the Ser CUG codons arose through individual mutations of various other codons, including universal Ser codons, by a nearly neutral process as described above. Indispensable Ser sites in a protein gene, only a single copy of which exists in an organism, would not have been involved directly in this type of reassignment, because a mutation, if it occurs at these sites, would be removed by negative selection. Therefore, such code changes could occur much more easily when multiples of the same gene exist.

6.6 Mitochondrial systems

In the following changes in the mitochondrial genetic code, production of unassigned codons would have been brought about by directional mutation pressure and/or by genomic economization. Unassigned codon capture could take place in one of the three ways outlined in Section 6.4. The mechanisms of the code changes to be described below are by no means decisive because of lack of enough information.

6.6.1 *UGA from stop to Trp*

The same or similar sequence of events as described for the *Mycoplasma* lineage probably occurred in the evolution of non-plant mitochondria. In the chlorophyte alga *Prototheca wickerhamii*, UGA would possibly be an unassigned codon (Wolff *et al.*, 1994). Perhaps, UGA has not been captured by Trp because $tRNA^{Trp}_{UCA}$ has not yet emerged (see Section 5.1). This would correspond to stage (b) in Fig. 6.18. It is possible that UGA Trp exists in mitochondrial species of chlorophyte algae other than *Prototheca wickerhamii.*

6.6.2 *CUN from Leu to Thr*

The reassignment of CUN codons from Leu to Thr during yeast mitochondrial evolution could have proceeded by the disappearance of CUN codons from the reading frames of mRNA, through mutation mainly to UUR Leu codons as a result of AT-pressure. This would have been accompanied by a loss of Leu-accepting ability of $tRNA^{Leu}_{UAG}$. This tRNA could have then acquired Thr-accepting activity through the appearance of an additional threonyl-tRNA synthetase. CUN codons that subsequently appeared from mutations of various other codons would have been translated as Thr (Osawa *et al.*, 1990*a*).

As filamentous fungi (CUN = Leu) and yeasts are phylogenetically close, the reassignment of CUN to Thr occurred in the yeast line after separation from their common ancestor. The base compositions and patterns of codon usage are strongly biased towards AT-base pairs in yeast and fungal mitochondria, indicating that these genomes are under AT-pressure. *Saccharomyces cerevisiae* mitochondria have a G + C content of only 18 per cent. The G + C content of codon silent sites is even lower, averaging less than 10 per cent. The preferential accumulation of A and T in genes results in a pattern of codon usage that favours A + T-rich synonymous codons. For example, the four yeast and fungal mitochondrial genomes employ Thr ACA/U codons almost exclusively over the synonymous Thr ACG/C codons. Under AT-pressure, the silent third positions of these codons would only rarely be expected to mutate to G or C, resulting in the observed accumulation of ACA/U codons. Similarly, AT-pressure would be expected to drive the accumulation of Leu UUA codons. As expected, UUR Leu codons (mostly UUA) predominate over CUN Leu codons (mostly CUA/U) in *Aspergillus* and *Neurospora* mitochondria. The ratio of UUR to CUN is about 14 : 1 in *Aspergillus* mitochondria. (In yeast mitochondria the number of Thr CUN codons remains small relative to UUR Leu.) (Table 6.5). It is thus reasonable to assume that an ancestor of yeast mitochondria used Leu CUN codons only rarely. The first position nucleotide of Leu CUR codons is silent in the genetic code of fungal mitochondria, and in all other known codes (except yeast mitochondria) and, presumably, it was silent in the ancestral code of yeast mitochondria. AT-pressure can convert Leu CUR codons directly to Leu UUR through silent substitution of the first position nucleotide. The first position of Leu CUY codons is not silent, but these codons may mutate silently to Leu UUR by mutating to CUR first, and then to UUR. Leu CUN codons in the ancestor of yeast mitochondria may have disappeared through silent mutations of Leu UUR. The fact that most Leu CUN sites in fungal mitochondria correspond to Leu UUR in yeast strongly suggests that this was the case.

The reassignment of codon CUN from Leu to Thr in yeast mitochondria was accompanied by the development of a new $tRNA^{Thr}$ that translates CUN codons. It was originally thought that this tRNA originated from mutations in the regular mitochondrial $tRNA^{Thr}_{UGU}$ (Li and Tzagoloff, 1979). Later, on the basis of sequence resemblance to *Neurospora crassa* mitochondrial $tRNA^{Leu}_{UAG}$, the yeast mitochondrial $tRNA^{Thr}_{UAG}$ was said to have originated from $tRNA^{Leu}_{UAG}$ (Sibler *et al.*, 1981). The reassignment of $tRNA_{UAG}$ from Leu to Thr requires a change in aminoacyl-tRNA synthetase recognition. It is likely that this change depended on the development of an additional threonyl-tRNA synthetase. A yeast nuclear gene (MST 1) coding for a mitochondrial threonyl-tRNA synthetase, which can aminoacylate only $tRNA^{Thr}_{UAG}$, has been identified (Pape and Tzagoloff, 1985). AT-pressure would have played a role in the conversion of $tRNA^{Leu}_{UAG}$ to $tRNA^{Thr}_{UAG}$ in yeast mitochondria. AT-pressure may affect the composition of tRNA genes in addition to influencing the frequency of their corresponding codons. *Saccharomyces cerevisiae* mitochondrial tRNA genes are relatively A + T-rich, averaging 33 per cent G + C. However, $tRNA^{Thr}_{UAG}$ is particularly enriched in A + T (79 per cent) (Osawa *et al.*, 1990*a*). The tRNA is unusual in another respect, i.e. in translating rare CUN codons. It is thus possible that, when AT-pressure causes CUN codons to fall into disuse in the ancestor of yeast mitochondria, the functional constraints on the corresponding $tRNA^{Leu}_{UAG}$ gene are thereby reduced, allowing it to accumulate mutations that would otherwise be deleterious. Because of the directional bias, most of these mutations are towards

Table 6.5 Yeast and fungal mitochondrial codon usage

	Yeast			Fungi	
	S. cerevisiae mitochondria	*T. glabrata* mitochondria		*A. nidulans* mitochondria	*N. crassa* mitochondria
Thr (CUA)	20	2	Leu (CUA)	13	24
Thr (CUC)	1	0	Leu (CUC)	0	2
Thr (CUG)	6	0	Leu (CUG)	2	4
Thr (CUU)	7	1	Leu (CUU)	13	23
Leu (UUA)	355	62	Leu (UUA)	395	112
Leu (UUG)	4	0	Leu (UUG)	4	11
Thr (ACA)	69	1	Thr (ACA)	72	16
Thr (ACC)	2	0	Thr (ACC)	0	0
Thr (ACG)	0	0	Thr (ACG)	0	1
Thr (ACU)	52	4	Thr (ACU)	51	37

Leucine and threonine codon usage is shown for 12 genes from *Saccharomyces* and *Aspergillus*, five genes from *Neurospora*, and two genes from *Torulopsis* mitochondria.
From Osawa *et al*. (1990*a*).

A+T, with the result that the G+C content of the tRNA gene decreases. Thus, unusual base composition may indicate that a relaxation of selective constraints has occurred. If a codon is retired from use entirely, the corresponding tRNA gene sequence will no longer be maintained by selective forces and may either be lost from the genome or become free to acquire a new function. Yeast mitochondrial $tRNA_{UAG}$ may have lost its ability to interact with leucyl-tRNA synthetase, and later acquired the ability to interact with MST 1, so that its gene was not removed from the genome. As CUN codons subsequently appeared in reading frames from mutations of various codons, they were 'captured' and translated as Thr by the 'new' $tRNA^{Thr}_{UAG}$ that evolved from $tRNA^{Leu}_{UAG}$. The change in aminoacylation of yeast mitochondrial $tRNA_{UAG}$ from Leu to Thr could have given rise to numerous Leu-to-Thr replacements in yeast mitochondrial proteins if Leu CUN codons had not previously disappeared. Such a change might have persisted if the number of CUN sites was so small that the mitochondrial proteins would not have been affected functionally by this change. If a concerted replacement of Leu by Thr had occurred at the time of reassignment, one would expect present-day yeast mitochondrial Thr CUN codons to correspond to Leu CUN in the homologous positions of fungal mitochondria.

Alternatively, the codon capture model proposes that Leu CUN codons disappeared prior to the reassignment and views present-day Thr CUN codons as being derived from various other codons. It predicts that present-day yeast mitochondrial Thr CUN codons should not correspond preferentially to Leu in fungal mitochondria. Out of 13 Thr CUN codons in yeast mitochondria, none corresponds to Leu CUN in *Aspergillus* mitochondria. Only one of nine yeast mitochondrial Thr CUN sites is occupied by Leu CUN in *Neurospora* mitochondria. The other yeast mitochondrial CUN sites are occupied in fungal mitochondria by Thr ACN, Val GUN, Ser UCN, and other codons. By comparison, the level of amino acid conservation at homologous yeast mitochondrial Thr ACN sites in *Aspergillus* and *Neurospora* is 48 per cent and 40 per cent, respectively. Furthermore, most Leu CUN sites in fungal mitochondria correspond to Leu UUR in yeast mitochondria in accordance with the suggestion that yeast mitochondrial Leu CUN codons disappeared prior to the code change.

Some of the present-day CUN Thr codons in yeast mitochondria could have been derived from ACN Thr. However, ACN cannot be converted to CUN by a single mutation. Presumably, some of the ACN codons first mutated to AUN (Ile/Met) or UUR (Leu) via UCN (Ser) and thence to CUR (Thr). Other CUN Thr codons in yeast mitochondria were derived from various codons for other amino acids. Indispensable ACN Thr sites, if any, would not have been involved in this reassignment. About 85 per cent of Thr in yeast mitochondrial proteins is coded by ACN, and only 15 per cent by CUN.

6.6.3 *AUA from Ile to Met*

The amino acid assignment of codon AUA differs in the mitochondria of different species. The model in Fig. 6.14 indicates that independent switching of

AUA Ile to Met took place twice—once in the yeast ancestor and once in the metazoan ancestor (Osawa *et al.*, 1989*b*; Jukes and Osawa, 1990).

The reassignment of codon AUA from Ile to Met during mitochondrial evolution may be summarized as follows (Osawa *et al.*, 1989*b*): AUA Ile codons would have mutated mainly to AUU Ile codons because of constraints from elimination of $tRNA^{Ile}$ with anticodon LAU. Later, $tRNA^{Met}_{CAU}$ would have changed so as to pair with both AUG and AUA. AUA codons, formed by mutations of other codons (including AUG) would have reappeared and would have been translated as Met. Reading of both AUA and AUG codons as Met seems to be performed by codon–anticodon pairing with $tRNA^{Met}$, in which C at the anticodon first position would be modified to f^5C (Moriya *et al.*, 1994).

Yeast mitochondria *Aspergillus* and *Neurospora* share a recent common ancestor with yeasts, and in these fungal mitochondria AUA is the most frequently used Ile codon, as might be expected from their high genomic A + T content. Many of the AUA Ile sites in *Aspergillus* and *Neurospora* mitochondrial genes are replaced by AUY (mostly AUU) Ile in yeast mitochondria. No Ile AUA sites in *Aspergillus* and *Neurospora* mitochondria correspond to AUA Met in yeast mitochondria.

Because the mitochondrial code has evolved from the eubacterial code, translation of AUA as Ile and AUG as Met in *Aspergillus* and *Neurospora* mitochondria may be performed by $tRNA^{Ile}_{LAU}$ and $tRNA^{Met}_{CAU}$, respectively. In fact, genes for three species of tRNAs with CAU anticodons have been found in *Aspergillus nidulans* mitochondria—the initiator $tRNA^{Met\text{-}f}$, the primary elongator $tRNA^{Met}$, and possibly $tRNA^{Ile}_{LAU}$ specific for codon AUA (Köchel *et al.*, 1981). On the other hand, yeast mitochondria have only elongator and initiator $tRNAs^{Met}$, both with the CAU anticodon. These facts suggest that $tRNA^{Ile}_{LAU}$ disappeared before or concomitantly with disappearance of AUA codons. However, the removal of this tRNA gene would not have occurred by a one-step disappearance, because this would have made the predominant Ile codon, AUA, immediately untranslatable, which would be disruptive. We must assume the gradual disappearance of the gene for this tRNA, possibly by genomic economization, whereas the gene for $tRNA^{Met}_{CAU}$ remained unchanged. During this time, the AUA Ile codons mutated to AUY Ile and other codons, because of the constraint imposed by shortage of the $tRNA_{LAU}$. This resulted in codon AUA becoming unassigned. Elongator $tRNA^{Met\text{-}e}_{CAU}$ then acquired f^5C modification so as to translate both AUA and AUG as Met. At this stage, AUA could appear in reading frames from mutations of various codons, including AUG Met, and would be translated by the altered $tRNA^{Met}_{CAU}$.

Metazoan mitochondria Metazoan mitochondria contain only one species of tRNA with anticodon CAU, presumably functioning for both initiation and elongation. This indicates that one species of tRNA with anticodon CAU was deleted in the early stage of metazoan mitochondrial evolution. It is difficult to

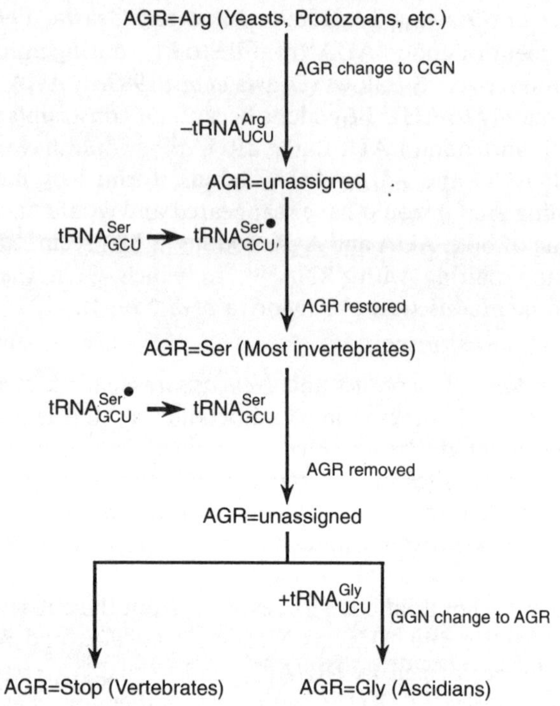

Fig. 6.22. Possible evolutional consequences of AGR codons in animal mitochondria. tRNA$^{Ser\bullet}_{GCU}$, that can translate all AGN codons as Ser, is derived from tRNA$^{Ser}_{GCU}$. tRNA$^{Ser\bullet}_{GCU}$: tRNA$^{Ser}_{GCU}$ that possibly received structural modification (see text) (from Yokobori *et al.* (1993), with modifications).

decide which tRNA^{Met} was deleted, but the reassignment of AUA from Ile to Met would have occurred in a way similar to that in yeast mitochondria.

6.6.4 *AGR from Arg to Ser, Gly, or stop*

The possible route of each code change may be summarized as follows (Fig. 6.22). During metazoan mitochondrial evolution, AGR codons became unassigned because of deletion of tRNA$_{UCU}$ and elimination of AGR codons by conversion to CGN Arg codons. Upon acquisition by tRNA$^{Ser}_{GCU}$ of pairing ability with AGR codons, some codons, mainly for Ser (and not for Arg) mutated to AGR, and were captured by tRNA$^{Ser}_{GCU}$ (Osawa *et al.*, 1989*a*). In the process leading to the ancestor of prochordates and vertebrates, the two AGR Ser codons were unassigned, presumably because of loss of pairing ability of tRNA$^{Ser}_{GCU}$ with AGR. In the prochordate line, the tRNA$^{Gly}_{CCU}$ gene duplicated and the anticodon of one of these mutated to UCU. AT-pressure caused some GGN codons to change to AGR, which were captured by tRNA$^{Gly}_{UCU}$ (Yokobori *et al.*, 1993). During vertebrate mitochondrial evolution, AGR stop codons were presumably created from UAG stop by deletion of the first nucleotide U and by use of R as the third nucleotide that existed next to the ancestral UAG stop (Osawa *et al.*, 1989*a*).

AGA is the most abundantly used Arg codon in mitochondria of simpler eukaryotes, presumably as a result of high AT-pressure, which is reflected in their high genomic A + T content. For example, the amino acid sequences deduced from several genes from mitochondria of the yeast, *Saccharomyces cerevisiae*, show that 86 per cent of Arg residues are coded by AGA, 4 per cent by AGG, and 10 per cent by CGN. To show the pattern of replacement of these Arg sites among various mitochondria, the AGR Arg sites in several genes of mitochondria from *S. cerevisiae* and a protozoan (*Trypanosoma*) were compared with the homologous sites of the corresponding mitochondrial genes in starfish, *Drosophila*, and five vertebrate species (Osawa *et al.*, 1989*a*). Most of the Arg (AGR) residues in yeast or *Trypanosoma* mitochondria are replaced by CGN Arg codons throughout metazoan mitochondria, in which AGR codons are used as either Ser or stop codons, suggesting that Arg at these sites is functionally important and so is well conserved. Thus, no direct replacements of AGR Arg by AGR Ser were revealed in the genes examined. Comparisons of AGR Ser sites in starfish or *Drosophila* mitochondrial genes with the corresponding sites in those from yeast or *Trypanosoma* reveal that AGR Ser sites in metazoan mitochondria were originally occupied largely by UCN Ser, AGY Ser, or codons for amino acids other than Arg.

In metazoan mitochondria, $tRNA^{Arg}_{UCU}$ for codons AGR is absent; this tRNA would have been deleted by reduction of genome size during evolution. The complete replacement of AGR, mostly by CGN Arg and a few other codons in metazoan mitochondria, suggests that AGR codons were converted mainly to CGN (a silent change) upon deletion of $tRNA^{Arg}_{UCU}$, so that AGR became unassigned.

Let us assume that the metazoan mitochondrial ancestor had a usage of AGR and CGN codons for Arg similar to that in yeast mitochondria (AGR : CGN = 9 : 1) before AGR became unassigned, because, like yeast mitochondria, genomes from simpler metazoan mitochondria are usually high in A + T. The process that led AGR to become unassigned may be explained using the following equations originally proposed by Sueoka for directional mutation pressure:

$$\hat{p} = \frac{v}{u+v} \text{ or } \frac{v}{u} = \frac{\hat{p}}{1-\hat{p}} \tag{6.1}$$

where v and u are defined as the effective conversion rates of A/T to G/C and G/C to A/T, respectively and $\hat{p}$ is the G + C content at equilibrium (Sueoka, 1962). For this specific example, let us define v and u as the effective conversion rates of AGR to CGN, and CGN to AGR, respectively, and $\hat{p}$ as the CGN content at equilibrium ($\hat{p} = 0.1$ in yeast mitochondria). Considering here the directional mutation pressure, D (μ_D of Sueoka, 1988) and the change in translational efficiency of tRNA, T, as two major factors to influence Arg codon usage, and defining $v = D_v \cdot T_v$ and $u = D_u \cdot T_u$, eqn (6.1) may be expressed as:

$$\hat{p} = \frac{D_v \cdot T_v}{D_v \cdot T_v + D_u \cdot T_u}$$

or

$$\frac{D_v \cdot T_v}{D_u \cdot T_u} = \frac{\hat{p}}{1 - \hat{p}}. \tag{6.2}$$

D_v and D_u are constant unless the genomic G+C contents change, whereas T_v and T_u are variable with the change of translation efficiency of codons by tRNA. The general formulas for T_v and T_u are not known, but T is equal to unity when tRNA availability is at saturation, and may move to zero along with the decrease in tRNA availability. In yeast, AGR is read by major $tRNA^{Arg}_{UCU}$, and CGN by minor $tRNA^{Arg}_{ACG}$, suggesting that T_u is a little higher than T_v. In the above argument, selective constraints on phenotype have not been treated as a major factor and are considered as neutral.

In the metazoan mitochondrial ancestor, AGR at the important Arg sites mutated to CGN Arg codons by completely selective constraints ($T_u = 0$) even in the presence of AT-pressure ($D_u > D_v$), because conversion of AGR to CGN was the only way to conserve these Arg residues. However, the removal of $tRNA^{Arg}_{UCU}$ would not have occurred by one-step deletion. If it had occurred, predominant AGR codons would have become untranslatable, which would certainly have been deleterious. What probably happened was that $tRNA^{Arg}_{UCU}$ gradually lost pairing ability with AGR codons ($T_u \rightarrow 0$), so that u decreased. Finally, the gene for this tRNA disappeared ($T_u = 0$), so that no more conversion of CGN to AGR could occur ($u = 0$). Gradual disappearance of a tRNA has been reported in one of the $tRNAs^{Trp}$ in *Mycoplasma* and one of the $tRNAs^{Pro}$ in liverwort chloroplast. By this time, all AGR codons had disappeared from reading frames, mainly by conversion to CGN ($\hat{p}$ becomes 1), and the tRNA for CGN codons would presumably have increased in amount. In this way, AGR became, without disruption, an unassigned codon pair available for subsequent capture by an amino acid.

Following this, the structure of $tRNA^{Ser}_{GCU}$, pairing primarily with AGY codons, was modified so that the altered $tRNA^{Ser}$ translated AGR codons (see below). AGR (mainly AGA) codons, pairing with this $tRNA^{Ser}$, then appeared in the reading frames by mutations of AGY Ser codons and other codons. AGR codons were thus captured by Ser.

The AGR codons that have been captured by Ser are translated either by $tRNA^{Ser}_{UCU}$ (in *Ascaris*) or by $tRNA^{Ser}_{GCU}$ (in other invertebrates). These tRNAs have undergone structural changes, bringing about the hypothetical G : R pairing between anticodon GCU and codons AGR in addition to AGY. The anticodon UCU in *Ascaris* mitochondria would have reappeared with a new assignment (Ser) by mutation of anticodon GCU of $tRNA^{Ser}$ under AT-pressure and seems to translate all AGN codons as Ser by 4-way wobbling (see p. 89 and p. 91).

The two 'new' Ser codons AGG and AGA should have been removed from reading frames in messenger RNA before these codons became either Gly (prochordates) or stop codons (vertebrates). Shortly before this, or in the early phase of prochordate/vertebrate mitochondrial evolution, $tRNA^{Ser}_{GCU}$ underwent further changes leading to abolition of pairing ability with codons AGR. AGR codons were presumably first removed by mutation, mainly to AGY, by strong selective constraints resulting from the loss of AGR translation by anticodon GCU, so that AGR codons became unassigned.

Throughout above descriptions, changes in the pairing ability of anticodon GCU ($G:Y \rightarrow G:N \rightarrow G:Y$) have been attributed to the structural changes of the $tRNA^{Ser}_{GCU}$ without experimental evidence. Further studies are obviously needed on this point.

As noted above, in addition to GGN, ascidian mitochondria use AGA and AGG as Gly codons. After prochordates had separated from the ancestors of vertebrates, the gene for $tRNA^{Gly}_{UCC}$ was duplicated and the anticodon of one of the duplicated $tRNA^{Gly}$ genes was converted from TCC to TCT because of AT-pressure (66 per cent A + T). At the same time, AT-pressure caused some GGN codons to change to AGR codons, which were captured by $tRNA^{Gly}_{UCU}$ transcribed from one of the duplicated $tRNA^{Gly}$ genes, resulting in the translation of AGR codons as Gly. The usual $tRNA^{Gly}_{UCC}$ for universal GGN Gly codons in animal mitochondria possesses unmodified uridine at the anticodon first position, which enables the anticodon UCC to pair with all four GGN Gly codons. Presumably, $tRNA^{Gly}_{UCU}$ has modified uridine at this position so as not to pair with AGY Ser codons. The gene for $tRNA^{Gly}_{UCU}$ reveals a high homology with that for $tRNA^{Gly}_{UCC}$ for universal Gly codons GGN, suggesting that duplication of the gene for $tRNA^{Gly}_{UCC}$ and mutation of one duplicate to $tRNA^{Gly}_{UCU}$ occurred (Yokobori *et al.*, 1993). Generally, gene duplication in the mitochondrial genomes is quite rare (if it happens at all), because of the strong tendency of reducing the genome size. However, it could happen as in the case of $tRNA^{Trp}$ gene duplication in the *Mycoplasma* line, in which genome economization is also evident.

During vertebrate mitochondrial evolution the unassigned AGA and AGG codons appeared as stop codons in vertebrate mitochondrial genomes. The simplest way to use unassigned AGR as stop codons would be for AGR to appear at the translation–termination site by a 1-step mutation from the pre-existing stop codon. However, the standard mitochondrial stop codons UAA and UAG cannot change to AGR by single nucleotide replacement. Two-step mutations, e.g. UAR $\rightarrow$ UGR $\rightarrow$ AGR result in a disappearance of the stop codon at the first mutation step, because UGR is not a stop codon and is read-through as Trp until the next stop codon appears, as a result of which a longer peptide is produced. This possibility is not very likely.

Four AGR stop codons are known in vertebrate mitochondrial genes: (1) AGA in human COI; (2) AGG in human ND 6; (3) AGA in bovine cytochrome b; and (4) AGA in *Xenopus* ND 6. The appearance of these AGR stops is explained in the following way (Osawa *et al.*, 1989*a*) (Fig. 6.23): The ancestor for the gene had used UAG as a stop codon with the sequence for termination region 5'-NNNUAGR-3'.

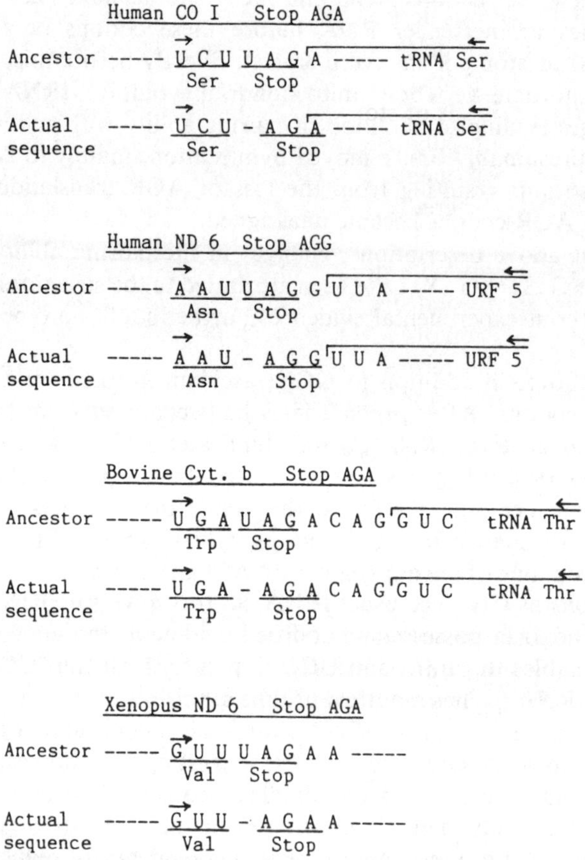

Fig. 6.23. Possible mechanism for conversion of UAG to AGR stop codons in vertebrate mitochondria. – denotes a deletion (from Osawa *et al.*, 1989*a*).

Deletion of U from the UAG stop by a 1-step mutation resulted in the present sequence 5′-NNNAGR-3′ without disruption of amino acid sequence or of position of the stop codon. This scheme can be applied to all of the known cases, as illustrated in Fig. 6.23. In the human mitochondria COI gene, the third nucleotide A of the AGA stop codon is complementary to the 3′-terminal U of the antisense sequence for the tRNA^{Ser} gene. Deletion of U from the ancestral UAG stop did not result either in change of amino acid sequence of COI or of the sequence of tRNA^{Ser}. In the case of *Xenopus*, another putative protein gene ND 5 is encoded within the opposite strand of DN 6 and their terminal regions are overlapped. The terminal sequence of ND 6 is 5′-UGCGUUAGAA-3′ and the antisense sequence of ND 6 is 5′-UUCUAACGCA-3′, where UAA is used as stop codon for ND 5. If we assume that the ancestral terminal sequence of ND 6 is 5′-UGCGUUUAGAA-3′, from which the present sequence 5′-UGCGUUAGAA-3′ resulted by deletion of U from UAG stop, the ancestral sequence of the antisense strand should be 5′-UUCUAAACGCA-3′. The present antisense sequence 5′-UUCUAACGCA-3′ was derived by deletion of A at the middle position of the ancestral UAA stop. This deletion produced a new stop codon—UAA—because the nucleotide next to the ancestral UAA stop is A. Thus, neither the amino acid sequences nor the positions of stop codons for the two genes were affected.

Another, but not mutually exclusive, possibility would be as follows: When the ancestral terminal sequence for a given gene was 5′-NNNUGRUAR-3′ or 5′-NNNGGRUAR-3′, where Trp UGR or Gly GGR was functionally dispensable, UGR or GGR could mutate to AGR stop by a single mutation. This produced a sequence with double stops as AGRUAR. UAR could then change its sequence, because the second stop UAR was no longer needed and was therefore free to mutate. Conversion of UGR, especially UGA, to AGR could occur, as the Trp codon UGA at the terminal region could be a vestigial remnant of the ancestral bacterial double stop UGAUAR and is functionally unimportant.

6.6.5 *AAA from Lys to Asn; AUA from Met to Ile as a reversal of Ile to Met; UAA from stop to Tyr*

The deduced amino acid assignments AUA for Ile and AAA for Asn in the mitochondria of echinoderms and platyhelminths, and the probable assignment of UAA for Tyr in planarian mitochondria, could have all resulted from similar codon captures (Osawa *et al.*, 1992). It is reasonable to speculate that, in the ancestral mitochondria of these animals, codons AUR Met and AAR Lys were translated by $tRNA^{Met}_{CAU}$ and $tRNA^{Lys}_{UUU\ (or\ CUU)}$, respectively, and UAA was a stop codon recognized by RF-1. The three codons AUA, AAA, and UAA would have disappeared from coding sequences following a gradual loss of function of the corresponding tRNAs and of the UAA-recognizing function of RF-1, respectively. The following two possibilities are suggested for loss of assignment of these three codons.

1. Ancestral mitochondrial $tRNA^{Lys}$ in echinoderms and platyhelminths, or $tRNA^{Met}$ in yeast and most metazoans, could formerly have had anticodon CUU or CUA capable of translating both AAA and AAG, or AUA and AUG, because some animal mitochondrial tRNAs with anticodon CNN (such as $tRNA^{Met}_{CAU}$ in mammalian and *Aedes* mitochondria and $tRNA^{Lys}_{CUU}$ in *Aedes* mitochondria), seem to recognize both codons NNA and NNG (Dubin and HsuChen, 1984; HsuChen *et al.*, 1983; Sprinzl *et al.*, 1989). Presumably, C is modified to, e.g. f^5C. The functions of Lys or Met tRNA changed gradually so as to translate only codon NNG, possibly by removal of the formyl residue from f^5C. In planarian mitochondria, RF (possibly RF-1) became specific for stop codon UAG (Bessho *et al.*, 1991). As a result, the NNA codons—AAA, AUA, or UAA—in coding frames would have been replaced by synonymous NNG codons and become unassigned.
2. GC-pressure in the genomes of the ancestors of these animal mitochondria resulted in mutation of the three NNA codons to NNG. Indeed, the G content of

Table 6.6 Changes in codons (*underlined*) while amino acid sequence remains constant

Original sequence	Asn ...	Lys ...	Lys ...	Asn ...	Lys ...	Asn ...	Lys ...	Asn
	AAC...	AAG...	AAA...	AAC...	AAG ...	AAU ...	AAA ...	AAU
	GUU	*CUU*	**UUU*	*GUU*	*CUU*	*GUU*	**UUU*	*GUU*
AAA, **UUU* disappear	AAC ...	AAG ...	AAG ...	AAC ...	AAG ...	AAC ...	AAG ...	AAC
	GUU	*CUU*	*CUU*	*GUU*	*CUU*	*GUU*	*CUU*	*GUU*
Anticodon change	<u>AAA</u> ...	AAG ...	<u>AAG</u> ...	AAC ...	AAG ...	<u>AAA</u> ...	<u>AAG</u> ...	AAU
	**GUU*	*CUU*	<u>*CUU*</u>	**GUU*	*CUU*	**GUU*	<u>*CUU*</u>	**GUU*
	Asn ...	Lys ...	Lys ...	Asn ...	Lys ...	Asn ...	Lys ...	Asn

Anticodons are shown in italic. Codon AAA for Lys disappears by converting to a synonymous Lys codon AAG. Anticodon **UUU* for codon AAA also disappears either by its removal from the genome or by mutation to CUU. These changes would occur either by GC-pressure or genomic economization exerted on the tRNA genes. Codon AAA then appears from mutation of AAC Asn and is captured by anticodon **GUU*, which was derived from *GUU*, to pair with codon AAA. *G would be inosine or equivalent that pairs with U, C, or G of the codon third nucleoside.
From Jukes and Osawa (1991) with modifications.

the codon third positions is markedly higher for echinoderms and *Fasciola hepatica* than for yeasts or *Drosophila* spp. Also, $tRNA^{Lys}_{UUU}$ for AAA and AAG codons changed to $tRNA^{Lys}_{CUU}$, which is specific to codon AAG. Similarly, $tRNA^{Met}_{CAU}$ has become specific to codon AUG and RF became specific to codon UAG in planarian mitochondria.

In either case, the NNA codon that disappeared was later captured by an amino acid that had previously been assigned only to an NNY 2-codon set. In echinoderm mitochondria, there is only one species of $tRNA^{Ile}$ (anticodon GAU in DNA) and $tRNA^{Asn}$ (anticodon GTT in DNA). These tRNAs must be responsible for reading NNA codons in addition to NNY. This could have followed a conversion of anticodon GNN to INN or equivalent, which would pair with all three codons NNU, NNC, and NNA for the same amino acid, AUY/A for Ile (echinoderms and planaria), AAY/A for Asn (echinoderms, *Fasciola*, and planaria), UAY/A for Tyr (planaria). A model for the code change from AAY Asn to AAA Asn is shown in Table 6.6.

These hypothesized changes in nuclear and mitochondrial codes all provide for the disappearance of a codon, followed by its reappearance with a different assignment. In almost every case, the amino acid sequences of the proteins involved are unchanged, so the reassignments are non-disruptive, but sometimes sequences are changed temporarily (yeast mitochondrial CUN Leu to Thr and *Candida* spp. CUG Leu to Ser).

7 Selenocysteine is coded by UGA

7.1 UGA as a selenocysteine codon

Selenium (Se) is an essential nutritional trace element in the diet of animals. Some selenoproteins, such as mammalian glutathione peroxidase and *E. coli* formate dehydrogenase, contain Se in the form of selenocysteine (Sec) at the active centre of the corresponding enzyme. Sec plays an essential role in maintaining a lower redox potential for these enzymes through the resulting selenol group as compared with the thiol group of Cys. This may be advantageous for anaerobic environments. In fact, selenite is required for the production of formate dehydrogenase activity in *E. coli*, which is expressed and is active under anaerobic conditions (Pinsent, 1954). Replacement of Sec by Cys results in a 75 per cent reduction in enzyme activity (Zinoni *et al.*, 1986).

In 1986, it was announced that Sec in *E. coli* formate dehydrogenase and a murine glutathione peroxidase was coded by the codon UGA (Zinoni *et al.*, 1986; Chambers *et al.*, 1986). Formate dehydrogenase containing Sec was also reported to occur in a methanogenic bacterium, *Methanococcus vannielii* (Shuber *et al.*, 1986). Se-containing dehydrogenases from some methanogenic bacteria and various sulphate-reducing anaerobic bacteria most probably contain Sec coded by UGA, as some of the genes for the dehydrogenases have been confirmed to contain an in-frame TGA codon. Glycine reductase selenoprotein A from a number of *Clostridium* spp. has been known to contain a Sec residue. UGA was found in the gene for this enzyme from *C. purinolyticum* at the position corresponding to location of the Sec residue in the gene product (see Stadtman, 1990; Burk and Hill, 1993). The gene for iodothyronine 5′ deiodinase (5′DI) from rat liver has TGA, which encodes Sec in the gene product (Berry *et al.*, 1991*a*). Remarkably, selenoprotein-P (Sel-P) of rat plasma contains eight Sec residues per mol of protein and in addition, this protein has a high content of Cys and His. Sel-P may serve as a protective agent against free radicals or other reactive molecules (Read *et al.*, 1990). Indeed, cDNA for both rat and human Sel-P contains ten TGA codons for Sec (Hill *et al.*, 1991).

Sec is inserted by a minor and specific tRNA with anticodon *UCA [$tRNA^{(Ser)Sec}$] that base-pairs with codon UGA (Leinfelder *et al.*, 1988). The distribution of Sec in nature may be examined by detecting Sec in proteins, Sec-coding UGA in genes, or by the occurrence of $tRNA^{(Ser)Sec}$ or its gene. The gene for $tRNA^{(Ser)Sec}$ apparently is ubiquitous in eubacteria as demonstrated by its presence in *Salmonella typhimurium*, *Proteus vulgaris*, two species of *Klebsiella*, *Serratia marcescens*, two species of *Enterobacter* (Heider *et al.*, 1989), *Clostridium thermoaceticum*, and *Desulfomicrobium baculatum* (Tormay *et al.*, 1994).

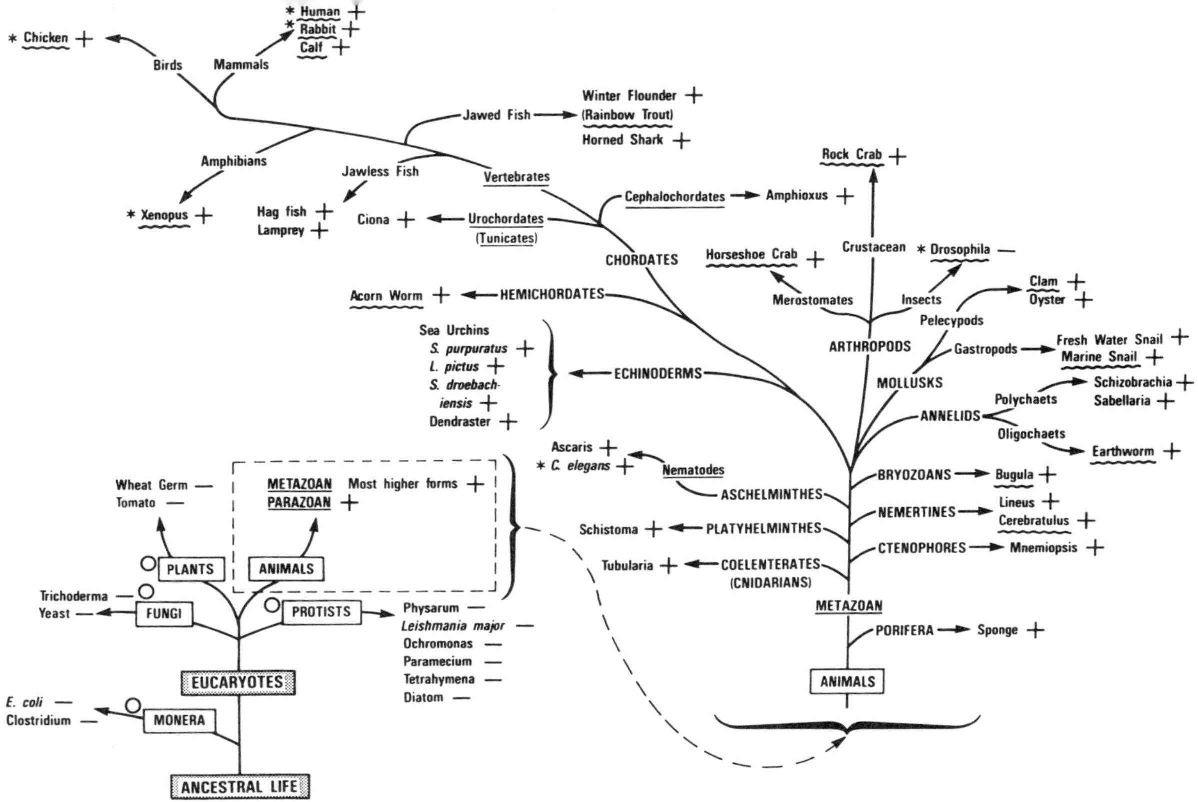

Fig. 7.1. Distribution of tRNA$^{[Ser]Sec}$ gene. Symbols designate the following: + positive response of restricted genomic DNA to the human tRNA$^{[Ser]Sec}$ gene probe; – no response; wavy underline, identification of the tRNA$^{[Ser]Sec}$ gene product within the seryl-tRNA population; * determination of the sequence of the tRNA$^{[Ser]Sec}$ gene; ○ presence of tRNA$^{[Ser]Sec}$ was demonstrated by other methods (from Lee *et al*. (1990), with addition of ○ symbols).

Lee *et al.* (1990) extensively studied the distribution of Sec by the use of the human tRNA$^{(Ser)Sec}$ gene as a probe to identify homologous sequences in the genomic DNAs from a wide variety of eukaryotes. As shown in Fig. 7.1, tRNA$^{(Ser)Sec}$ gene was detected throughout animal kingdom. tRNA$^{(Ser)Sec}$ also exists in a higher plant (*Beta vulgaris*), in a filamentous fungus (*Glioeladium virens*) and in two widely divergent protists (Hatfield *et al.*, 1992). It may be concluded that tRNA$^{(Ser)Sec}$, and thus the use of UGA as a codon for Sec, is widespread, if not ubiquitous, in nature. Neither selenoproteins containing Sec nor tRNA$^{(Ser)Sec}$ have been reported to occur in organelles.

7.2 Pathway of selenocysteine biosynthesis

Böck and his co-workers (Böck *et al.*, 1991; Forchhammer and Böck, 1991) have demonstrated that four genes were required for insertion of Sec into protein in prokaryotes. They are:

Sel C. *Sel C* is the structural gene for tRNA$^{(Ser)Sec}$. tRNA$^{(Ser)Sec}$ is aminoacylated with Ser by seryl-tRNA synthetase and the efficiency of aminoacylation is much less than the canonical tRNA^{Ser} (Förster *et al.*, 1990).

Sel A. The product of the *Sel A* gene is Sec synthetase. It converts seryl-tRNA$^{(Ser)Sec}$ to selenocysteyl-tRNA$^{(Ser)Sec}$ in two steps. The first step involves the dehydration of serine to form a complex between the enzyme and an acrylyl-tRNA intermediate (Forchhammer and Böck, 1991).

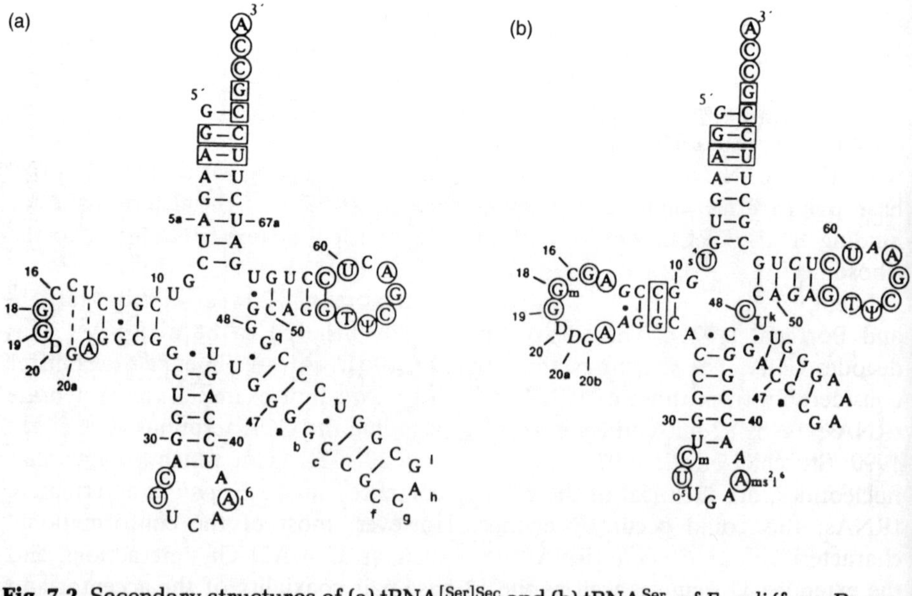

Fig. 7.2. Secondary structures of (a) tRNA$^{[Ser]Sec}_{UCA}$ and (b) tRNA$^{Ser}_{UGA}$ of *E. coli* (from Baron *et al.*, 1993).

Sel D. The *Sel D* product, with Mg^{2+}, ATP, and selenide then catalyses the 2,3 addition of H_2Se to the acryl double bond, producing selenocysteyl-$tRNA^{(Ser)Sec}$ (Leinfelder *et al*., 1990).

Sel B. The product of this gene is a specific elongation factor (designated SELB) that specifically participates in the insertion of Sec from Sec-$tRNA^{(Ser)Sec}$ into polypeptide in place of EFTu (Forchhammer *et al*., 1989, 1990; Baron and Böck, 1991). SELB is clearly distinct from EFTu.

Thus, it is apparent that the biosynthesis of Sec attached to $tRNA^{(Ser)Sec}$ occurs in a unique way as compared with the usual aminoacylation of other tRNAs.

7.3 Structure of selenocysteine tRNA

The structure of *E. coli* $tRNA^{(Ser)Sec}$ has been studied extensively by Baron *et al.* (1993). $tRNA^{(Ser)Sec}$ is 95 nucleotides long and contains a much longer extra-arm than that of $tRNA^{Ser}$ (Fig. 7.2). There are also a number of other features distinct from those of the canonical $tRNA^{Ser}$. $tRNA^{(Ser)Sec}$ is aminoacylated with Ser by seryl-tRNA synthetase so that the serine-identity determinants must exist in both $tRNA^{(Ser)Sec}$ and $tRNA^{Ser}$. In *E. coli*, eight determinants sites have been identified in $tRNA^{Ser}$. They are located at the end of the acceptor stem and in the second base-pair of the D-stem. The six determinants in the acceptor stem are conserved in $tRNA^{(Ser)Sec}$. Baron *et al.* (1993) have stated that 'the absence of CII-G24 identity pair, as well as conformational differences at the level of extra-arm junction to the body, likely explains the 100-fold decrease in charging efficiency of $tRNA^{Sec}$ [$=tRNA^{(Ser)Sec}$] as compared to $tRNA^{Ser}$'. The peculiar features of the sequence of $tRNA^{(Ser)Sec}$ makes a novel set of tertiary interactions possible, notably U14-A21-G8, and C16-C59 (Fig. 7.3). Baron *et al.* (1993) suggest that these conformational characteristics provide 'the highest specificity for interaction with selenocysteine synthase'. An eight base-pair acceptor stem (rather than seven for $tRNA^{Ser}$) forms 13 base pair coaxial stacking with a T-stem of five base pairs (rather than four in an L-shape in tRNA), is the determinant for binding to the SELB protein that carries the selenocysteyl-$tRNA^{(Ser)Sec}$ to the ribosome.

In eukyaryotes, $tRNA^{(Ser)Sec}$ was discovered more than 20 years ago by Hatfield and Portugal (1970). The tRNA was initially believed to be a $tRNA^{Ser}$ that decodes UGA. The structures and sequences of $tRNA^{(Ser)Sec}$ in eukaryotes differ considerably from those of $tRNA^{(Ser)Sec}$ in *E. coli*. For example, the vertebrate $tRNA^{(Ser)Sec}$ is 87 nucleotides in length, excluding the CCA terminus (Lee *et al.*, 1990; Sturchler *et al.*, 1993) (Fig. 7.4). Only 29 sites, 11 of which are invariant nucleotides, are identical in the 87 sites compared in the *E. coli* and vertebrate tRNAs; this could occur by chance. However, most of the conformational characteristics in *E. coli* $tRNA^{(Ser)Sec}$, such as U14-A21-G8 interactions and the extended D-stem, as well as the 13 base-pair coaxiality of the acceptor and T-stem, are conserved in *Xenopus* $tRNA^{(Ser)Sec}$ (Sturchler *et al.*, 1993). The

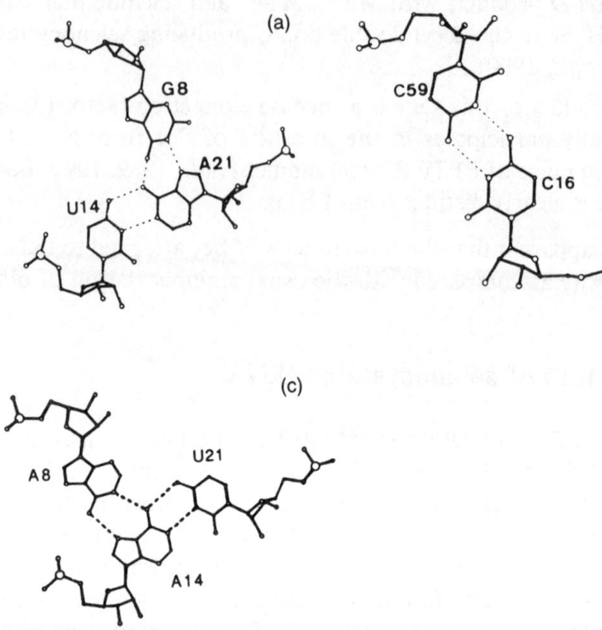

Fig. 7.3. The novel tertiary interactions in the structure tRNA$^{[Ser]Sec}$. (a) The U14-A21-G8 triple interaction and (b) the C16-C59 interaction in *E. coli* tRNA$^{[Ser]Sec}$ (from Baron *et al.*, 1993); (c) the A14-U21-A8 triple interaction in *Xenopus* tRNA$^{[Ser]Sec}$ (from Sturchler *et al.*, 1993).

resemblances between prokaryotic and eukaryotic tRNA$^{(Ser)Sec}$ would suggest a common origin, although it is possible, but less likely, that the two tRNA$^{(Ser)Sec}$ groups have undergone a kind of convergent evolution with regard to function in forming the present-day set of tRNAs.

Ohama *et al.* (1994) showed that the discriminator base (G73) of human tRNA^{Ser} and tRNA$^{(Ser)Sec}$ has an absolute requirement for seryl-tRNA synthetase recognition in the sequence-specific manner. (In contrast, the discriminator base is said to be not critical in *E. coli* tRNA^{Ser} (Asahara *et al.*, 1994).) The extra-long arm domain is also a critical element in the Ser-identity for both tRNAs. The recognition of this region in tRNA^{Ser} does not seem to be sequence-specific and appears to have a more global effect in the identity, such as length-specific effects. On the other hand, G : C pairings in this region of tRNA$^{(Ser)Sec}$ play a sequence-specific and an orientation-specific role in the Ser-identity, as mutational conversion of G : C to C : G pairs greatly diminishes the serylation of this tRNA. The identity elements in tRNA$^{(Ser)Sec}$ responsible for conversion of seryl-tRNA$^{(Ser)Sec}$ to selenocysteyl-tRNA$^{(Ser)Sec}$ have not been identified conclusively in either prokaryotic or eukaryotic tRNA$^{(Ser)Sec}$.

Very little is known about the pathway of selenocysteyl tRNA$^{(Ser)Sec}$ synthesis in the eukaryotic system.

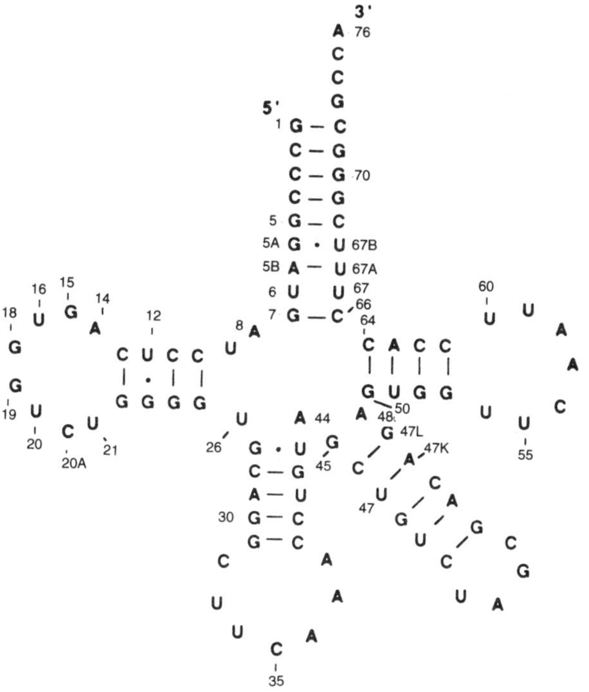

Fig. 7.4. Secondary structure of *Xenopus* tRNA$^{[Ser]Sec}_{UCA}$ (from Sturchler *et al.*, 1993).

7.4 Translational context of selenocysteine UGA

The efficient translation of UGA as Sec in the mRNA of *E. coli* formate dehydrogenase requires the presence and identity of 40 bases 3′ to the internal UGA (Zinoni *et al.*, 1990). These bases form a hairpin-loop next to the UGA codon. A similar structure was also found in mRNAs for *Enterobacter aerogenes* formate dehydrogenase and *Desulfovibrio baculatus* [NiFeSe] hydrogenase (Zinoni *et al.*, 1990) (Fig. 7.5). Deletions within this loop that destroy the hairpin structure prevent UGA translation (Baron *et al.*, 1990). The secondary structure is a binding site for a unique C-terminal domain of the SELB protein, suggesting that the complex of selenocysteyl-tRNA$^{(Ser)Sec}$ and SELB recognizes the site of insertion.

In contrast to prokaryotes, stem-loop forming sequences in the 3′ untranslated region (3′UTR) are required for incorporation of Sec at UGA sites in eukaryotes (Berry *et al.*, 1991*b*; Berry *et al.*, 1993; see also Farabaugh, 1993). In 3′UTR, such a stem-loop was found in both the rat and human genes for type 1 5′ID in which one UGA codon is read as Sec. A similar stem-loop structure was predicted in the 3′UTR for glutathione peroxidase mRNA. The loop-forming sequence in the 3′UTR (SECIS element) is well-conserved (Fig. 7.6). Both rat and human Sel-P genes each contain ten internal TGA codons for Sec. In two cases, two TGA

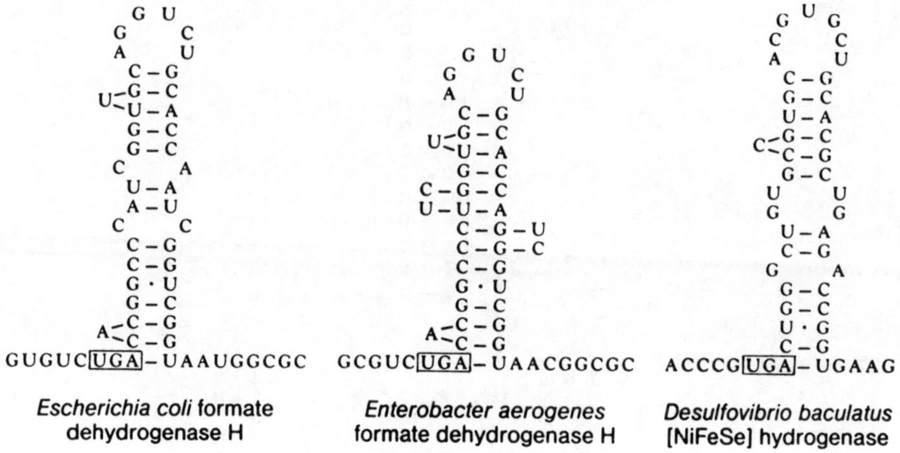

Fig. 7.5. Predicted structures in the vicinity of UGA codon in mRNAs coding for Sec in proteins (from Zinoni *et al.*, 1986).

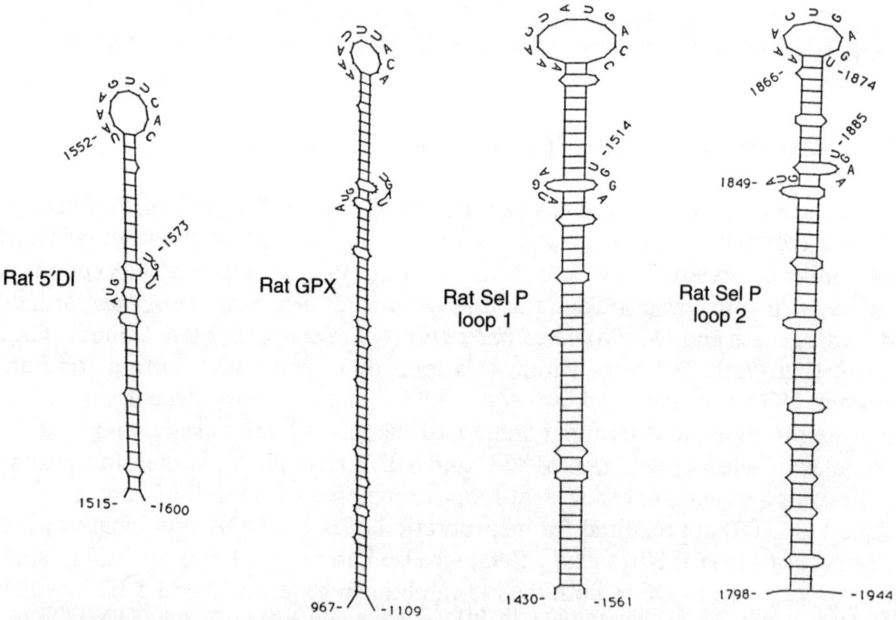

Fig. 7.6. Predicted secondary structures in SECIS elements in the 3′ untranslated regions from rat 5′DI, rat glutathione peroxidase (PGX) and rat Sel P stem-loops 1 and 2 (from Berry *et al.*, 1993).

codons are separated by a single codon. Hill and co-workers (1991) found that 'determination of the possible folding patterns' in the coding region of mRNA 'predicted a stable-loop structure following only 1 of the 10 UGAs'. Thus, recognition of internal UGA as Sec in eukaryotes is not dependent on the secondary structure near the UGA sites. Moreover, the gene for Sel-P has only two stem-loop sequences in the 3′ non-coding region, and therefore 'there must not be a one-to-one functional connection between stem-loops and insertion site'. Remarkably, UGA codons randomly inserted into the 5′ID gene are translated efficiently as Sec (see Farabaugh, 1993).

A model has been proposed to account for the recognition of UGA codons as Sec in eukaryotic systems (Berry *et al.*, 1993; see also Burk and Hill, 1993) (Fig. 7.7). Selenocysteyl tRNA$^{(Ser)Sec}$ would interact with a special elongation factor (similar to SELB). This complex would then recognize the specific stem-loop structure—SECIS—in the 3′UTR, so that selenocysteyl tRNA$^{(Ser)Sec}$ would have access to the UGA codons. Thus, all the UGA codons may be read by anticodon *UCA of this tRNA regardless of their positions in mRNA. Such a mechanism cannot operate in prokaryotes because transcription and translation are coupled. Eukaryotic selenoprotein mRNAs do not use UGA as a stop codon (e.g. UAG is the stop codon for rat Sel-P; and UAA for human Sel-P, human 5′ID, and rat glutathione peroxidase), perhaps to avoid mistranslation of the stop codon as Sec. mRNAs other than those for selenoproteins do not have the specific stem-loop structure in the 3′UTR responding to the complex of selenocysteyl tRNA$^{(Ser)Sec}$ and SELB-like elongation factor, so that UGA can be used as a stop codon.

Generally, a codon cannot have more than one meaning simultaneously. UGA is a remarkable exception; in-frame UGA at the specific site in mRNA is read as Sec, and UGA at the termination site functions as a stop codon. In *Mycoplasma capricolum*, UGA is commonly used as a Trp codon; it remains to be seen if UGA is used as a Sec codon at certain sites in mRNA.

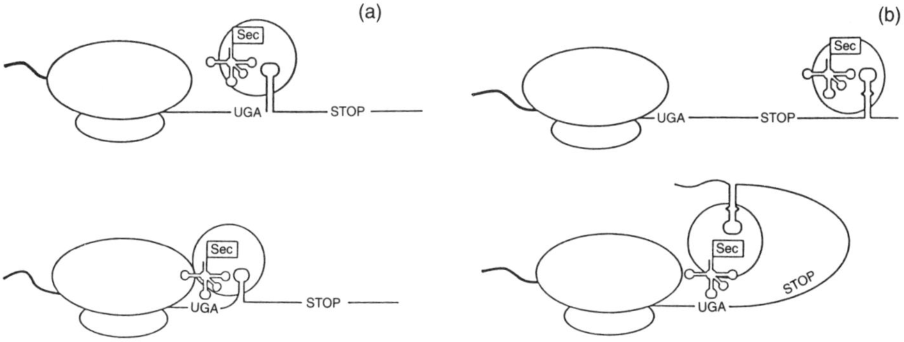

Fig. 7.7. Model for selenocysteine incorporation in prokaryotes and eukaryotes. (a) Prokaryote. Selenocysteyl-tRNA$^{[Ser]Sec}$-SELB complex recognizes the stem-loop structure distal to UGA on mRNA. The ribosome moves to the codon UGA. (b) Eukaryote. The complex recognizes SECIS element and anticodon of selenocysteyl-tRNA$^{[Ser]Sec}$ pairs with UGA at any site in mRNA (from Berry *et al.*, 1993).

7.5 Evolution of UGA-coded selenocysteine synthesizing system

Sec has been sometimes called the 21st amino acid (Söll, 1988; Böck *et al.*, 1991), because the protein synthesizing system can use Sec in addition to the 'magic 20'. The different non-universal genetic codes described in Chapter 6 have all been derived from the universal genetic code. Whether UGA Sec is derived as a recent evolutionary improvement, or is a relic in the 'anaerobic world' is still under much discussion. Ubiquitous occurrence of the Sec-insertion system throughout present-day organisms suggests that the system arose before divergence of prokaryotes and eukaryotes. However, this does not distinguish between these two possibilities.

Leinfelder and co-workers (1988) suggest that 'UGA was originally a codon for Sec in the anaerobic world, perhaps two to three billion years ago, and after introduction of oxygen into biosphere this highly oxidizable amino acid could be maintained only in anaerobic organisms or in aerobic systems which evolved special protective mechanisms'. In the aerobic world, nearly all Sec residues in protein were switched to Cys. Codon UGA could have 'acquired other functions such as its more familiar role in termination' while being retained for rare use in coding for Sec.

The UGA site for Sec in formate dehydrogenase of *E. coli* is substituted by UGC for Cys in *Methanobacterium formicicum* (Shuber *et al.*, 1986). The specific activity of the *E. coli* enzyme is four to five times higher than that of *Methanobacterium*. This suggests that Sec and Cys are counterparts and the Cys-containing enzyme was derived from the Sec-containing one, or vice versa (Böck and Stadtman, 1988).

Evolutionary change of UGA from Sec to a stop codon, or to Cys UGY is rather hard to trace. Jukes (1990) suggests that UGA could not have changed abruptly from coding for Sec to a stop codon. In the anaerobic world, the UGN family box was assigned to both Cys and Sec with anticodon UCA. When oxygen entered the biosphere, nearly all the Sec was switched to Cys, which retained UGN codons and anticodon UCA. This anticodon duplicated, and one duplicate mutated to GCA; the present Cys anticodon paired with UGY. The other UCA anticodon was captured by a 'new amino acid'—Trp—with codon UGR. GC-pressure changed anticodon UCA to CCA, pairing only with UGG, the present Trp codon, and UGA disappeared, except for rare use in coding for Sec.

If this series of events is true, we must suppose that there was once an elaborate system similar to that which exists today for introduction of Sec into protein. This is rather unlikely, especially as $tRNA^{(Ser)Sec}$ is probably a derivative of $tRNA^{Ser}$ and not of $tRNA^{Cys}$. Jukes *et al.* (in Osawa *et al.*, 1992) then proposed the opposite point of view—that the Sec system is a 'sophisticated evolutionary innovation . . . rather than the survival of primitive systems'. The more primitive forms of the genes for the dehydrogenases, etc. used Cys codon UGY at the present Sec UGA sites. At this stage of the code, perhaps corresponding to the early code (see Section 9.2), UGA and UGG were codons for Trp. Note that Sec synthetase differs from other 'magic 20' aminoacyl-tRNA synthases: It functions in modification of Ser rather than in direct attachment of Sec to $tRNA^{(Ser)Sec}$. This would favour the view that Sec arrived after the establishment of the code for 20 amino acids. When the Sec-incorporation system first appeared, it would have been simpler than the present form. Presumably, appearance of a primitive form of Sec-synthetase came first, followed by the emergence of $tRNA^{(Ser)Sec}$ from $tRNA^{Ser}$. As the use of Sec would have become more advantageous for anaerobic organisms than the use of Cys, the system could have undergone evolutionary refinement by positive selection. This would have involved the conversion of UGY to UGA, restriction of UGA sites in mRNA so as not to be recognized by tryptophanyl-tRNA (see below), and the development of SELB from EF-TU to enable recognition of the internal UGA sites by $tRNA^{(Ser)Sec}$. It is significant that Sec incorporation occurs when UGA in *E. coli* formate dehydrogenase is replaced by Cys codons UGU or UGC (Baron *et al.*, 1990). This would suggest that UGY codons were used for both Cys and Sec in early evolution. Mutation of UGY to UGA would have resulted in avoidance of competition with selenocysteyl-$tRNA^{(Ser)Sec}$, with cysteyl-$tRNA^{Cys}$ for decoding, so that the internal UGA has become used exclusively for Sec. Another problem is the possible dual use of UGA as a Sec codon and a Trp codon in *Mycoplasma*. Whether UGA is used for Sec in this bacterium, as in other bacteria, so that the selenocysteyl–$tRNA^{(Ser)Sec}$–SELB complex does not compete with tryptophanyl-tRNA with anticodon UmCA remains to be determined.

Why was UGA not captured by Sec with elimination of UGA as a stop codon as in the case of Trp UGA in *Mycoplasma capricolum* or in mitochondria? Perhaps, because Sec was used only rarely, nature had to give UGA an optional function—Sec coding—in addition to the termination function.

8 RNA editing

8.1 Biological meaning of RNA editing

RNA editing can also be called 'nucleotide substitution'—insertions or deletions made in the RNA sequence. Typically, deleterious changes that have occurred in DNA are restored at the RNA level by this process. The origins of the known forms of RNA editing in various organismic lineages are probably independent.

8.2 Green plant mitochondria

8.2.1 *Is CGG for Trp in plant mitochondria?*

Originally it was suggested that the genetic code in green plant mitochondria differed from the universal code by the use of codon CGG for Trp instead of Arg. Maize, wheat, and *Oenothera* mitochondrial genes sometimes encode CGG triplets at the well-conserved Trp sites, TGR, in mitochondria from other organisms (Fox and Leaver, 1981; Hiesel and Brennicke, 1983; Schuster and Brennicke, 1985; Stern *et al.*, 1986; Fox, 1987). However, no $tRNA^{Trp}$ with anticodon CCG for codon CGG has been found in plant mitochondria (Maréchal *et al.*, 1985). Plant mitochondria are known to use UGG for Trp, and only one $tRNA^{Trp}$ with anticodon CmCA for codon UGG was found. It was also frequently observed that green plant mitochondrial genes possess certain codons leading to deleterious amino acid replacement, such as Cys TGT by Arg CGT when these are translated as such (Covello and Gray, 1989; Gualberto *et al.*, 1989).

These dilemmas were solved by the unexpected discovery of RNA editing, which came to light when gene sequences were compared with the corresponding RNA sequences (and later with the cDNA sequences). For example, C in the codon CGG (sites 87 and 129) in wheat COXII genes, which corresponds to Trp TGG or TGA in other mitochondria, is replaced by U in the RNA sequence, indicating that the apparent 'non-universal' Trp codon CGG at the DNA level is edited to the universal Trp codon UGG at the RNA level (Covello and Gray, 1989; Gualberto *et al.*, 1989). The same is true for other 'deleterious' codons in DNA. It is now evident that green plant mitochondria use the universal genetic code.

Table 8.1 summarizes RNA editing in green plant mitochondria. The C-to-U substitution is often seen, whereas the U-to-C substitution is quite rare. The editing sites are predominantly in the coding regions, with a few in non-coding regions and ribosomal DNA (Hiesel *et al.*, 1989; Yang and Mulligan, 1989; Conklin *et al.*, 1991; Knoop *et al.*, 1991; Wissinger *et al.*, 1991; Binder *et al.*,

1992). Most of the editing in the coding regions is at the replacement sites, although some occurs at silent sites (Table 8.2). Many codons found to be edited in some sites were not edited in many other sites that did not need editing. Some sites in COXII are edited differently in different plant species (Covello and Gray, 1990). So, RNA editing is highly selective. This type of editing is called 'substitution editing' (Bass, 1992), and is said to be a result of deamination of C.

Editing involving G or A has not been reported in plant mitochondria. Perhaps RNA editing operates only for T–C, because of a probable lack of editing mechanism for G or A.

Interestingly, liverwort mitochondria do not seem to edit their mRNA in the manner used by angiospermous plant mitochondria. Notably, all four Trp sites in the liverwort COXII gene are TGG; no CGG Trp sites have been found. In every case, the amino acids and the termination signals deduced from 'unedited codons' in liverwort mitochondria are identical to those specified by edited codons in wheat and evening primrose mitochondria (Ohyama *et al.*, 1991). These observations lead us to speculate the origin and evolution of RNA editing.

There are at least two possibilities. One is that RNA editing is an ancient mechanism in the RNA world and liverwort mitochondria discarded it during evolution. The second is that RNA editing evolved after the emergence of the angiosperms. Ohyama *et al.* (1991) and Covello and Gray (1993) are inclined to believe the latter view.

8.2.2 *A model for the evolution of RNA editing*

Covello and Gray (1993) proposed a 3-step model for the evolution of RNA editing, which consists of: (1) the appearance of RNA editing activity (assuming that the editing is of recent origin); (2) mutation at editable nucleotide positions and fixation by (genetic) drift; and (3) maintenance of RNA editing activity by natural (positive) selection. The following discussion is based mainly on their proposal with some explanations and comments.

The appearance of RNA editing The first step involved a few neutral mutations in the pre-existing cytidine deaminase, and their fixation by genetic drift, so that the enzyme began to act on RNA to change some Cs to Us. When RNA editing activity first appeared, its specificity would not have been absolute. Full RNA editing activity would have resulted in the editing of pre-existing Cs in dispensable as well as indispensable positions. For example, the COXII gene of liverwort mitochondria, which lacks RNA editing, contains about 15 per cent Cs among total nucleotides, 10 per cent of which occupy the first and second positions of codons (Ohyama *et al.*, 1991). This would mean that 10 per cent of amino acids would be replaced by other amino acids if the RNA editing activity affected every C. The consequences of this would be extremely deleterious. It is therefore more likely that the editing activity was low so as to produce a mixture of differently edited transcripts. Some inactive transcripts from a gene receiving C-to-U edition at the essential sites may be overcome by the rest of the active transcripts

Table 8.1 Examples of RNA editing occurring in plant mitochondria

Triplet codons				Species							
DNA	(amino acid)	mRNA	(amino acid)	Wheat	(Ref.)	Oenothera	(Ref.)	Maize	(Ref.)	Pea	(Ref.)
CTT	(Leu)	UUU	(Phe)	+	(1, 2)						
TCT	(Ser)	UUU	(Phe)	+	(2, 3)			+	(12)		
TCC	(Ser)	UUC	(Phe)	+	(1, 2)	+	(7)				
TTC	(Phe)	UUU	(Phe)					+	(12)		
CTC	(Leu)	UUC	(Phe)	+	(3, 4)	+	(7)	+	(12)		
CCC	(Pro)	UUC	(Phe)			+	(8)				
TCA	(Ser)	UUA	(Leu)	+	(5, 6, 2, 4)	+	(8, 9)	+	(12)	+	(12)
CCA	(Pro)	UUA	(Leu)					+	(12)		
CTG	(Leu)	UUG	(Leu)	+	(4, 6)						
TCG	(Ser)	UUG	(Leu)	+	(3, 5)	+	(8)	+	(12)		
CCA	(Pro)	CUA	(Leu)	+	(2, 3, 4)	+	(9)	+	(12)		
CUC	(Leu)	CUU	(Leu)			+	(10)				
CCC	(Pro)	CUC	(Leu)			+	(10)				
CCG	(Pro)	UUG	(Leu)	+	(2)						
CCT	(Pro)	CUU	(Leu)	+	(1, 2, 5)	+	(10)	+	(12)	+	(12)
GUC	(Val)	GUU	(Val)	+	(4, 6)						
CAC	(His)	UAC	(Tyr)	+	(2, 3)	+	(11)				
CAT	(His)	UAU	(Tyr)	+	(3)	+	(11)				
ACT	(Thr)	ACC	(Thr)	+	(1)						
ATT	(Ile)	ACU	(Thr)			+	(10)				
CGG	(Arg)	UGG	(Trp)	+	(2, 3, 5)	+	(8, 10)	+	(12)	+	(12)
CGA	(Arg)	UGA	(Stop)	+	(4, 6)	+	(9)				
ACG	(Thr)	AUG	(Met)	+	(5)			+	(12)	+	(12)
CGT	(Arg)	UGU	(Cys)	+	(2, 5)						
CCT	(Pro)	UCU	(Ser)			+	(8, 10)			+	(12)
CCC	(Pro)	UCC	(Ser)	+	(2)			+	(12)		
AGC	(Ser)	AGU	(Ser)							+	(12)
ACT	(Thr)	AUU	(Ile)	+	(2)						
ACA	(Thr)	AUA	(Ile)	+	(2)	+	(10)				

from the same gene, in which only Cs at the dispensable sites are edited to U. The outcome of these changes will be neutral or slightly disadvantageous, so that the editing activity is easily lost unless the RNA editing subsequently acquires an advantageous role.

Before the second step is discussed, the advantageous role of RNA editing is introduced by describing the effects of its absence.

In the absence of RNA editing, a T-to-C mutation at a single essential T site is deleterious and is removed by negative selection. Back-mutation of C to T is not possible. This type of negative selection occurs in other organelles and organisms when the RNA editing is absent.

In the presence of RNA editing, a T-to-C mutation at the essential T site can occur as a neutral change, because C is edited to U at the RNA level. At the same time, back-mutation of C to T becomes possible, as the C is not eliminated by negative selection. In effect, the nucleotide at such a site in the transcript is U, regardless of whether the site was occupied by C or T in DNA. For example, there are two synonymous codons for Phe in the genetic code, and yet the numbers of the synonymous codons encoded in DNA increase 4-fold in the presence of RNA editing, i.e. CTT, CCT, CTC, CCC, TCT, and TCC, in addition to TTT and TTC, provided that a C of the first and/or second codon positions is at the editable site. When arising from mutation(s) of TTT or TTC, these codons can be all edited to UUU or UUC in messenger RNA. Therefore, RNA editing is advantageous for the rescue of negative selection caused by T-to-C mutation.

Mutation at editable positions Even in plant mitochondria, mutations other than T-to-C are eliminated by negative selection, because of the absence of other RNA editing activity.

When RNA editing activity first emerged, only one T site (or at most a few) would have mutated to C, and been edited to U. It is most unlikely that multiple editable C sites were created from Ts, because mutations usually occur one by one, and at intervals. If a single, or even a few, editable sites remain for a long time, the editing activity will be lost when the site is T instead of C, such that RNA editing is not needed.

Maintenance of RNA editing activity The next step is the maintenance of RNA editing activity. Covello and Gray (1993) state:

> If we invoke the law of large number [of editable C sites], the situation becomes very different, and the process is now essentially irreversible. . . . many of which [Cs at editable sites] would be deleterious if encoded and expressed as C. . . . The process becomes

---

+ RNA editing occurs in mitochondria of the indicated plant. Amino acids in parentheses indicate if triplets in DNA or mRNA are translated as such. Editing occurs mostly on the replacement sites. The resultant amino acid assignments of mRNA codons coincide to the consensus ones among mitochondria from other organisms in most cases; see also Table 8.2. References cited: 1, Gualberto *et al.* (1990); 2, Lamattina and Grienenberger (1991); 3, Gualberto *et al.* (1989); 4, Nowak and Kuck (1990); 5, Covello and Gray (1989); 6, Begu *et al.* (1990); 7, Schuster *et al.* (1990*b*); 8, Hiesel *et al.* (1989); 9, Schuster and Brennicke (1990); 10, Schuster *et al.* (1990*a*); 11, Wissinger *et al.* (1991); 12, Covello and Gray (1990).

Table 8.2 RNA editing sites in plant COXII coding regions

Species	No. of editing sites			Percentage of amino acids changed
	Total	Silent	Non-silent	
Wheat	17	1	16+	5.8
Maize	20++	2	18+	6.5
Pea	13	1	12	4.6

+ includes two editing sites in a codon; ++ includes partial editing in two codons.
From Covello and Gray (1990), with modifications.

co-evolutionary, in the sense that RNA editing activity affects the evolution of editable sites. . . . The higher the number of 'functionally important' editable positions, the lower the probability that in course of evolution these positions would all 'revert' to T at the DNA level [if appropriate evolutionary pressure exists; see below], such that editing would not be required. This leads to maintenance of RNA editing by natural [positive] selection. In this way we envisage that a functionally redundant process could arise and be maintained within a genetic system in a co-evolutionary manner.

From the above discussion, it is apparent that RNA editing activity must have been kept after the very first T-to-C mutation(s) had occurred until the subsequent mutations took place, so that the numbers of editable C sites increased. Two evolutionary events may be responsible for ensuring an accumulation of editable sites: (1) the acceleration of mutation rate; and (2) the predominant occurrence of T-to-C mutations over C-to-T. GC-directional mutation pressure seems to meet these requirements. With increasing GC-pressure, the rate of T-to-C mutation increases and, at the same time, C-to-T reversion by genetic drift becomes harder and harder. In its extreme form C-to-T reversion practically stops, so that RNA editing becomes the only way to maintain the essential Us in RNA.

The genomic G+C content of plant mitochondria is considerably higher than that of non-plant mitochondria, suggesting that some GC-pressure has operated in the course of the evolution of plant mitochondria. Thus, an accumulation of Cs as a law of multiple editable sites could be greatly influenced by GC-pressure in the presence of RNA editing activity during evolution. It is probable that the RNA editing activity co-evolved, both quantitatively and qualitatively, with an accumulation of editable Cs in DNA as a result of predominant T-to-C conversions over C-to-T by GC-pressure. Naturally, GC-pressure brings about an increase in Gs. These Gs at the essential sites are eliminated by negative selection.

If the above speculations are correct, C-to-U RNA editing activity could be lost under AT-pressure, because T-to-C mutations hardly occur. An apparent lack of RNA editing in liverwort mitochondria might be because the genomic G+C content of this mitochondria is relatively low and therefore C-to-T RNA editing would not be so advantageous that such a system did not develop. Covello and Gray's model assumes:

> Mutations at editable positions generating identical RNA sequences are neutral. This assumption implies that RNA editing sites should be fairly evenly distributed among protein-coding and non-coding regions of mRNA.

However, most of the editable sites exist at the replacement sites of codons, and Cs at silent sites are scarcely edited. It would not be possible for the editing to occur at random and for only mRNA that is properly edited to function, because the population of such 'correctly edited' mRNA, if there was any at all, would be very small among a mixture of differently edited mRNAs with large possible combinations. Thus, the specificity of the present-day RNA editing must be highly stringent. The RNA editing system would have evolved so as to avoid edition of 'unchangeable Cs' (deleterious if edited) as well as Cs at silent positions. Covello and Gray (1993) argue:

> If there are large number of editing sites, the overall editing process may represent a significant cost in energy and/or efficiency of gene expression, which would result in an overall disadvantage. . . . The outcome might be that an RNA editing activity would tend to alter in specificity to avoid 'non-essential' sites while maintaining editing at sites where it is required for proper gene expression.

This argument is acceptable. An alternative view, which is not mutually exclusive from the above, would be as follows: The fact that the editable essential site can be either T or C, because of the presence of RNA editing, is evidently more advantageous than if the site must be only T in the absence of RNA editing, because there is less chance of negative selection. Therefore, the higher the number of essential editable sites, the better, so these sites are positively selected. However, neither of these arguments explains how RNA editing enzyme recognizes the proper C sites in mRNA. In fact, 'neither primary consensus sequence nor secondary structure could fully explain the specificity of editing' (Covello and Gray, 1990).

As described in Chapter 6, when an unassigned codon is reassigned, the codon appears with the emergence of a new tRNA that translates the codon. Subsequent increase in numbers of the reassigned codon is accelerated by directional mutation pressure, with an increase in the amount of the tRNA (and copy number of its gene) in a co-evolutionary manner. Thus, the evolution of RNA editing activity, as postulated above, resembles the evolution of the non-universal genetic code.

8.3 Chloroplasts

C-to-U editing, i.e. AUG (initiation codon) from ACG has been reported for maize (Hoch *et al.* 1991), tobacco, and spinach (Kudla *et al.*, 1992).

8.4 Eukaryotic nuclear systems

The gene for apoB 100 protein in mammalian nuclear system contains a Gln (CAA) codon (nucleotide 6666 of human cDNA). C-to-U substitution editing occurs under various physiological conditions at the CAA site to form a stop codon UAA, resulting in the production of apoB-48 protein (Chen *et al.*, 1987; Powell *et al.*, 1987).

A-to-G editing occurs in nuclear encoded mRNAs for two classes of brain glutamate receptors (Sommer *et al.*, 1991).

8.5 Mitochondria

Until recently, there were only a few cases of RNA editing other than mRNAs. However, editing systems that act on tRNAs were found in 1993. In the mitochondria of *Acanthoamoeba castellanii*, an amoeboid protozoan, the genes or $tRNA^{Met\text{-}f}$, $tRNA^{Ala}$, $tRNA^{Asp}$, and $tRNA^{Met2}$, which consist of a cluster together with the gene for $tRNA^{Pro}$, differ in sequence from their corresponding tRNAs. The changes are T to A, T to G, and A to G in the acceptor stem. These changes correct mismatched base-pairs in DNA, i.e. A : C to G : C (one case), U : C to G : C (three cases) and U : U to A : U (two cases). U : G and G : U pairs, which are often present in helical regions of RNA, are not edited (Lonergan and Gray, 1993).

Another example of tRNA editing is even more interesting. In marsupial liver mitochondria, the anticodon sequence of the gene for $tRNA^{Asp}$ is GCC, which, as such, would recognize two of the Gly codons GGY; the cDNA sequence is GTC, which would recognize GAY Asp codons (Janke and Pääbo, 1993).

As in the case of mRNA editing, the tRNA editing illustrated above shows that at least some deleterious mutations are not selected against, and are restored at the RNA level. One must therefore be cautious when inferring the tRNA sequences from the corresponding tRNA, especially in mitochondria. In fact, in many cases only the DNA sequence has been reported.

8.6 Insertion/deletion editing

Insertion/deletion editing has been detected in protozoan kinetoplasts and in the mitochondria of the slime mould *Physarum polycephalum*. The former exclusively involve insertion and deletion of uridine residues, whereas the latter involves only insertion of C. This form of editing will not be described in detail here, because codons are completed by the editing rather than being changed in meaning (for further discussion of this topic, see Bass, 1992).

9 Origin and early evolution of the genetic code

In general, it may be best to keep theories of evolution of the code as close as possible to existing experimental findings and flexible enough to accommodate future discoveries. It is unlikely that some mathematical solution will be found to explain, in a single tour de force, the nature of the code, although efforts to achieve this are numerous.

Thomas H. Jukes (1992)

. . . we assume that the modern translation apparatus is a vastly improved version of the original apparatus, but does not differ from it fundamentally. We make this assumption because molecular evolution is generally *conservative*, modifying and adapting old pieces of molecular machinery whenever possible, instead of devising totally new ones.

Alan M. Weiner (1987)

9.1 So many theories

The early evolution of the genetic code has been the subject of speculation. Theories to explain it are numerous; some of these are aesthetically pleasing but cannot be verified. This chapter endeavours to discuss those with some experimental support.

It is well-known that contemporary protein synthesis proceeds with participation of tRNA, aminoacyl tRNA synthetase (ARS), mRNA, ribosome, a considerable number of proteinous factors, 20 amino acids, ATP, GTP, etc. More than 120 species of RNAs and proteins are involved in the process, which is undoubtedly the most complicated biochemical reaction in the present-day organisms. The origin and early evolution of the genetic code cannot be discussed without considering the evolution of these components—especially tRNA, ARS, mRNA, and ribosomes—as these components must have co-evolved with the genetic code. The origin of amino acids is also important.

9.1.1 *Abiotic synthesis of amino acids*

Amino acids can be produced by electric discharge in a mixture of inorganic compounds—such as CH_4, NH_3, H_2O, and H_2. These compounds form a strongly reducing environment, which mimics the primitive atmosphere of the earth. The amino acids produced by electric discharge are very similar to those found in the Murchison meteorite. Miller (1987) states: 'It is extremely gratifying to see that such synthesis really did take place on the present body of the meteorite, and so it

Table 9.1 Molar ratios of some amino acids formed by abiotic synthesis

Amino acid	Molar ratio in amino acid from: Electrical discharge	Murchison meteorite
Glycine	100.0	100, 98
Alanine	180.0	36, 44
α-Amino-*n*-butyric acid	61.0	19, 18
Norvaline	14.0	14, 3
Valine	4.4	19, 10
Norleucine	1.4	2
Leucine	2.6	4
Isoleucine	1.1	4
Alloisoleucine	1.2	4
Proline	0.3	22, 16
Aspartic acid	7.7	13, 5
Glutamic acid	1.7	20, 18
Serine	1.1	

From Cronin and Pizzarello (1983) and Weber and Miller (1981); Reprinted from Osawa *et al.* (1992).

becomes plausible but not proved that they took place on the primitive earth.' The proteinous amino acids produced by electric discharge under various conditions are: Gly, Ala, Val, Leu, Ile, Pro, Asp, Glu, Ser, and Thr, with several non-proteinous amino acids such as α-amino-*n*-butyric acid, norvaline, norleucine, etc. (Table 9.1). Thus, there are at least ten proteinous amino acids produced by abiotic synthesis in the universal genetic code. It is not unreasonable to speculate that the primitive coding system could have used these amino acids. Gln and Asn are unstable and are not detected in the abiotic mixtures. Gln is formed by amidation of Glu-$tRNA^{Gln}$ in several organisms (see Section 9.1.9). Therefore Gln, and perhaps Asn, could have been introduced into protein synthesis by such a mechanism, even though they are not formed by abiotic synthesis. It is possible that some, if not all, of the other proteinous amino acids could have been similarly produced. It is important to point out that many proteinous amino acids can be produced abiotically without the biosynthetic pathways (see below), so that they can be incorporated directly into the primitive coding system.

α-Amino-*n*-butyric acid, nonvaline, and norleucine, which are commonly produced by abiotic syntheses, are absent from the universal genetic code. Weber and Miller (1981) speculate that these amino acids would have been coded for and subsequently discarded.

9.1.2 *Two different aspects of the evolution of the genetic code*

There are at least two different aspects to the treatment of origin of the genetic code (Orgel and Crick, 1993). One is the stereochemical aspect, which has sometimes been used to explain the nature and origin of the genetic code. The theory postulates the specific physicochemical interaction of amino acids with codons (Woese, 1969; Fig. 9.1(a)) or tRNA anticodons (e.g. Wong, 1988; Fig. 9.1(b)) such that, for instance, 'an amino acid and an anticodon of comparative hydrophobicity might interact directly with one another'. In an extreme form of stereochemical theory a lock-and-key relationship was postulated between the tertiary structure of the anticodon–discriminator base complex and each of the 20 amino acid species (Shimizu, 1982). All the stereochemical theories imply a fixed relationship of amino acids with codons or anticodons, so at least some codon assignments are predetermined by the structural constraints; this also occurs if an anticodon is the primary tRNA identity determinant (Fig. 9.1(c)).

The second theory does not depend upon a mechanistic fixation of amino acids to codons or anticodons. Instead, it states that the specific aminoacylation of tRNAs is independent from codons or anticodons, so that the amino acid assignments of codons in the genetic code were brought about accidentally (Weiner, 1987; Orgel and Crick, 1993; Schimmel *et al.*, 1993) (Fig. 9.1(d,e)).

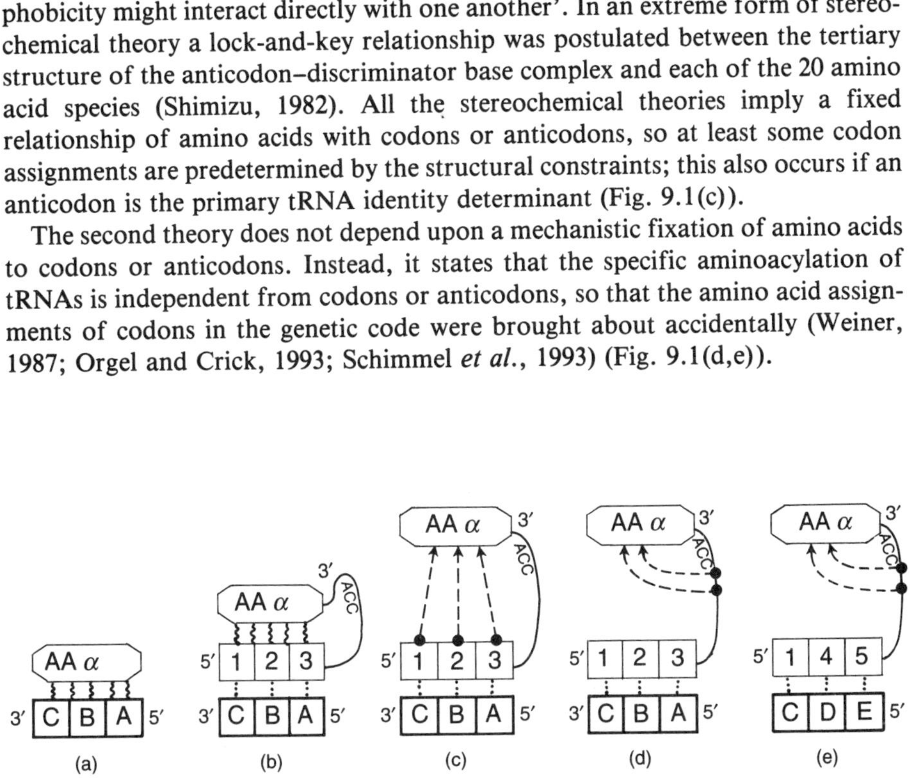

Fig. 9.1. Stereochemical and accidental determination of an amino acid assignment of a codon. (a) A codon A-B-C (5′→3′) fits amino acid α by physicochemical interaction, so that assignment of the codon for amino acid α is automatically specified. (b) Anticodon 1-2-3 (5′→3′) directly and specifically interacts with amino acid α. The result is the same as (a). (c) Anticodon 1-2-3 does not directly interact with amino acid α, but primarily determines the specificity to amino acid α through its interaction with ARS. The result is the same as (a) or (b). (d) Identity determinants for amino acid α are located outside the anticodon 1-2-3, so that the anticodon does not participate in specifying amino acid α. (e) If the anticodon sequence was 1-4-5 instead of 1-2-3 at the time of amino acid assignment of a codon, E-D-C (5′→3′), instead of A-B-C, could become a codon for amino acid α. For simplicity, the tRNA is shown as the anticodon plus 3′ half of the molecule including the CCA terminus.

9.1.3 *Stereochemical explanation of the origin of the genetic code*

Lewin (1974) argued that some sort of stereochemical fit between amino acids and codons is inevitable as the base for the primitive code, and that the products of translation evolve gradually to the present translation apparatus. 'It is otherwise difficult to see how a code which depends on its own translation products could have evolved'. He then speculated that, initially, polypeptides were formed by the stereochemical fit of amino acids to a polynucleotide template. The peptide and polynucleotide of this complex enabled each other to replicate in an autocatalytic cycle. Subsequently their relation became unidirectional and some of the polypeptides formed in this way might have evolved catalytic activities to improve their own production process, as well as that of other proteins. The present-day protein-synthesizing system might have evolved from such a complex, keeping the relationship between triplets and amino acids that first defined the original polypeptides. A reciprocally assisted replication of a polypeptide–nucleotide complex is highly speculative and its mechanism is hard to realize from present knowledge. It is also curious that Lewin made no mention on the involvement of tRNA in his model.

Stereochemical theories of the evolution of the genetic code seem to be no longer tenable in view of the lack of specific interaction of amino acids with

Table 9.2 The effect of suppressor anticodons introduced into tRNAs on specific aminoacylation of these tRNAs with cognate aminoacyl-tRNA synthetase

Retained	Diminished
$tRNA^{Ala}$	$tRNA^{Asp}$
$tRNA^{Arg}$	$tRNA^{Ile}$
$tRNA^{Cys}$	$tRNA^{Met}$
$tRNA^{Gln}$	$tRNA^{Thr}$
$tRNA^{Glu}$	$tRNA^{Val}$
$tRNA^{Gly}$	
$tRNA^{His}$	
$tRNA^{Leu}$	
$tRNA^{Lys}$	
$tRNA^{Phe}$	
$tRNA^{Pro}$	
$tRNA^{Ser}$	
$tRNA^{Trp}$	
$tRNA^{Tyr}$	

Respective tRNAs with CUA and UCA anticodons that pair with UAG and UGA stop codons, respectively, were aminoacylated by cognate aminoacyl-tRNA synthetase.
From McClain (1993).

codons (Crick, 1968; however, see Yarus, 1991; and Section 9.1.7), and the number of examples in which an amino acid acquired a different anticodon during evolution. For example in *Candida* spp. the 'universal' Leu anticodon is used for Ser. Also, an amino acid can be translated by experimentally modified anticodons in tRNAs. Normanly and Abelson (1989) synthesized or genetically altered the tRNA genes corresponding to 20 amino acids by substituting amber suppressor anticodon CUA for the anticodon. They found that 11 out of the 20 tRNAs could tolerate CUA anticodon without misacylating, although others were partly or wholly mischared. Their results, together with others, are summarized in Table 9.2. Hou and Schimmel (1988) succeeded in making *E. coli* $tRNA^{Phe}$ or $tRNA^{Cys}$ to accept Ala by introducing a G-C pair in the acceptor stem (the identity element of $tRNA^{Ala}$) into the analogous position of the respective tRNAs, without changing anticodons. Indeed, it is now evident that anticodon is not involved in the aminoacylation of at least some tRNAs, such as Ser, Leu, and Ala (Schimmel *et al.*, 1993). This is the most serious objection to the amino-acid–anticodon stereochemical relationship. Note, however, that in the present-day system the acceptance of amino acid by tRNA is mediated by interaction with the corresponding ARS; but that the role of ARS is obscure in the lock-and-key mechanism. It is easy to suggest an evasive argument, e.g. some stereochemical mechanism operated in the primordial coding system. Indeed, explanations as to how it was replaced by the present-day mechanism during evolution are all highly speculative.

9.1.4 *RRY and RNY hypotheses*

Lewin (1974) adopted a stereochemical explanation of the origin of the genetic code, but there is the dilemma of how the genetic code could have originated and how the most primitive protein synthesis could have arisen without the participation of code-dependent and highly specialized products, such as the ribosomes and ARSs.

Crick *et al.* (1976) proposed a model for primitive protein synthesis in relation to the origin of the genetic code. They postulated that the primitive mRNA could have been composed exclusively of RRY sequences. The alternating RRY sequences can be read in only one of the three possible frames without fixing the initiation codon (revival of a commaless code!). They suggested that the translation was carried out by tRNAs with anticodon loop sequence 3′...UGYYRUU...5′. This anticodon loop—tRNA positions 32 to 38—consists of the sequence 5′...YU-anticodon-RN...3′ in all tRNAs. This type of translation could have taken place before ribosomes emerged. The YYR sequence in the anticodon loop represents anticodons including UUR, UCR, CUR, and CCR pairing with codons AAY (Asn), AGY (Ser), GAY (Asp), and GGY (Gly), respectively. AAY codons in the primitive system could be Lys rather than Asn. Ser, Asp, and Gly are synthesized abiotically. These amino acids which, together with Lys, contain one basic and one acidic amino acid, are good candidates for the primitive protein (Jukes, 1977*b*). As ribosomes are not involved in this translation, Crick *et al.*

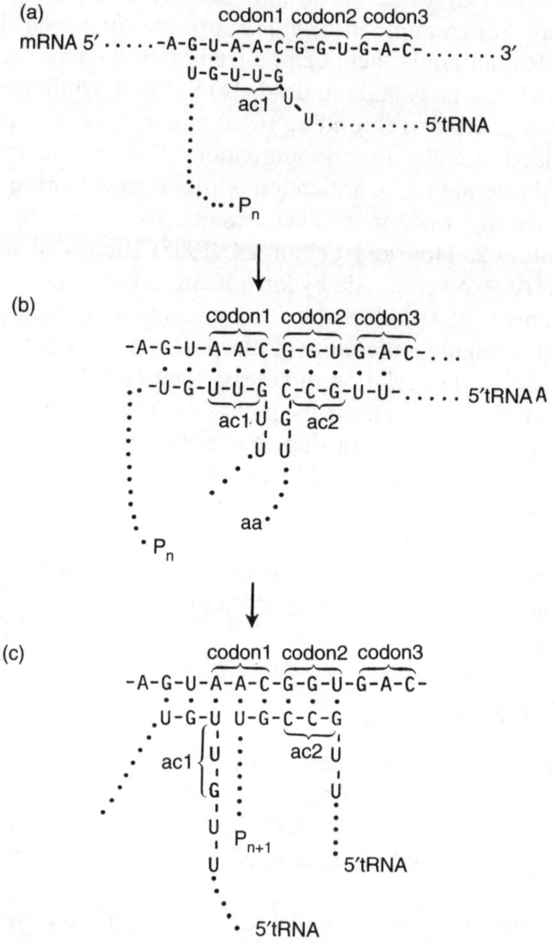

Fig. 9.2. Translation of a messenger RNA consisting of repeated RRY sequences. (a) mRNA and tRNA are interacted by the conventional codon-anticodon pairing with additional pairing between two bases (5′GU3′) adjoining the codon 1 (AAC Asn in this case) and two bases (3′UG5′) next to the third base of the anticodon 1. (b) The anticodon 2 (GCC) of the next tRNA pairs with the next codon 2 (GGU Gly). (c) A peptide bond is then formed between Pn and amino acid to form Pn + 1. This causes a conformational change in the two RNAs so as to make the anticodon loops 'twist'. Through this action, two bases (3′UG5′) adjoining the anticodon 2 displace the two bases (3′UG5′) in the anticodon 1. At the same time, two bases (UU) adjoining the first base (G) of the anticodon 2 become unpaired, just like translocation in the polypeptide formation on the ribosome. The first discharged tRNA pairing with only three bases, is released from mRNA. The codon 3 (GAC) is then translated in the same manner as described above. ac, anticodon; Pn, polypeptide (from Crick *et al.* (1976), and redrawn from Jukes (1977*b*), with modifications).

postulate that interaction, for the sake of stabilization, between messenger and anticodon involves five bases each using neighbouring 3′UG5′ or 3′UU5′ of anticodon in addition to the anticodon three bases. Translation proceeds by a flip mechanism involving conformational changes of the tRNA, as illustrated in Fig. 9.2. The peptides synthesized in this way could have been made up to the ribosome with primordial ribosomal RNA. The genetic code could then have expanded to include more amino acid to use the ribosome.

Inspired by the model of Crick *et al.*, Eigen and Schuster (1978) proposed that the first RNA involved in translation could have had a dual role as mRNA and tRNA. The structure of such an RNA could have been 5′...GNCGNCGNC...3′ (+strand of RNA) (but more probably, according to the authors, was 5′... GGCGCCGGC...3′). This could have been translated by tRNA with complementary anticodons CNG (−strand). Translation mechanisms are similar to those of Crick *et al.* (1976). In this model, mRNA–tRNA pairing is carried out by four bases each. This alternating GNC expanded to include GNY and then RNY, which provide for eight amino acids (Ala, Asn, Asp, Gly, Ser, Thr, Tyr, and Val); most of these amino acids are formed abiotically. Further expansion could have been possible only when a ribosomal translation system developed (for details, see Eigen and Schuster, 1978).

As RNY is a self-complementary sequence, primitive RNA could have acted both as mRNA and tRNA, as proposed by Eigen and Schuster (1978). This implies that protein synthesis was initially carried out by a single type of RNA molecule. Throughout the RRY and RNY hypotheses mentioned above, the role of ARS in the primitive protein synthesis is entirely obscure. How was the primitive tRNA aminoacylated?

Shepherd (1984) observed that RNY codons predominate over YNN in present-day DNA sequences, so that RNY in coding sequences could be a residuum of the primitive message. Analyses based on rates of silent substitutions, frequencies of nucleotide doublets, and synonymous codon usage ruled out support for a primitive RNY gene structure (Wong and Cedergren, 1986). RNN codons predominate over YNN, first because Glu, Asp, and Lys are all RNN and all exist in proteins at levels higher than their representation in the genetic code table (Jukes *et al.*, 1975), and second because the stop codons are YNN and occur only once in each gene for a protein. This predominance of RNN over YNN is therefore related to protein function and chain termination rather than to a residue of a primitive code (Jukes in Osawa *et al.*, 1992).

9.1.5 *RNA code*

The reading of a codon in the contemporary protein synthesizing system is performed on the ribosome by specific pairing with the corresponding anticodon in the aminoacylated tRNA.

Schimmel *et al.* (1993) proposed a scheme of evolution of the genetic code from the operational RNA code described in Section 2.6.1 (Fig. 9.3). This hypothesis weighs heavily on the notion that the genetic code is a historical accident. In its

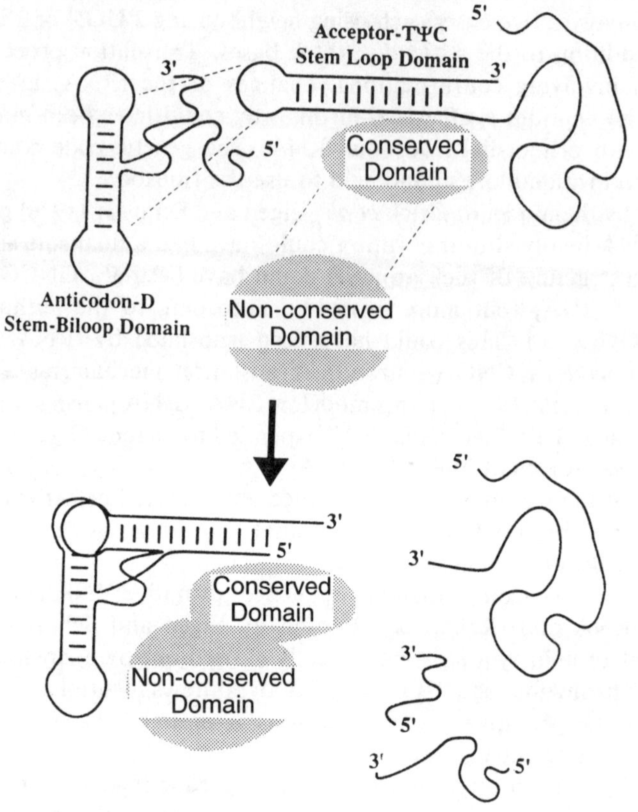

Fig. 9.3. A model for the assembly of ARS–tRNA complex. The interactions of the conserved synthetase domain with an RNA acceptor domain results in an operational RNA code for amino acids. The addition of the anticodon-D stem-biloop domain is co-ordinated by the addition of a non-conserved synthetase domain. As the full tRNA and ARS are assembled, amino acids become associated with specific anticodon and the genetic code is thereby established. It is possible that the early RNA domains were parts of larger RNA parts, which may have been ultimately discarded (from Schimmel *et al.*, 1993).

essence, the primordial 'conserved domain' of ARS was small and interacted with the amino acid attachment site of RNA, resulting in aminoacylation of the RNA. 'Aminoacylated RNA molecules might be bound together on an RNA matrix [precursor of (r)RNA] and react spontaneously to yield peptides'. With the development of an RNA template-directed, ribosome-dependent process, the anticodon domain was joined with the hairpin stem-loop, so as to be able to interact with the template by anticodon. The significance of the 'random' peptides synthesized independently from the template is not clear. Alternatively, the aminoacylated hairpin stem-loop could have had some function other than peptide formation. Schimmel *et al.* do not comment on how the template-dependent process arose. The addition of the non-conserved domain of ARS to the conserved domain would have taken place at this stage. Presumably, selective forces acted to increase the accuracy and efficiency of translation, so that the tRNA molecules and ARS may have undergone further differentiation, possibly in a co-evolutionary manner, and eventually have reached the contemporary L-shaped tRNA and ARS. The important aspect of this scheme would be that 'the particular algorithm of the genetic code might, therefore, be accidental' (i.e. not predetermined), resulting from the quasi-random combination of anticodons in the distal domain with amino-acid-specific acceptor helices. In this scheme, Schimmel *et al.* postulated the 2-base genetic code for 'a limited number of amino acids', followed by subsequent addition of the third base and more amino acids, which led to the present genetic code. The insertion of one base next to the pre-existing duplet code at regular intervals is unlikely to have occurred. Whatever its mechanisms, misinsertion of one base between the first and second bases causes a hopeless disturbance in the previous amino acid sequences of proteins, and would be lethal. The code is most likely to have started as a triplet. It is, however, possible that in the ancient code the third letter of codon did not participate in specifying amino acids (see p. 151). Another problem is how would the primordial synthetase have developed before the emergence of the template-dependent protein synthesizing system. According to Schimmel *et al.*'s scheme, ARSs are 'among the oldest proteins on the planet'. It is difficult to imagine how the specific ARSs depending on the genetic code could have emerged before its development. Did substrate RNA interact with some adequate peptides ('pre-genetic code proteins') among those synthesized non-enzymatically? This possibility is unlikely, as pointed out by Lewin (1974), because there is no way of transferring the specific amino acid sequences of these peptides to the offspring. This problem will be discussed in Section 9.1.6.

9.1.6 *Ribozyme origin of the genetic code*

The chick-and-egg relationship between the genetic code and its products may be solved, or at least becomes explainable, by the discovery of ribozyme (Cech, 1986). Taking advantage of the biological functions of ribozyme, Weiner (1987) and Maizels and Weiner (1987) proposed a model to explain the evolution of whole translational elements, including the origin of the genetic code (Fig. 9.4). In short, tRNAs, ARS, ribosome, the genetic code, and mRNA evolved in that order. A population of relatively homogeneous tRNA, or a single tRNA, was formed from the tRNA-like structure of RNA genomes. The first (single) ARS specific for a particular kind of basic amino acid, such as Lys, arose from RNA replicase, which depended on the replicase for replication. The first ribosome might have been a part of ARS itself, and participated in the formation of only short Lys peptides with lysyl-tRNA by translating a 'built-in-template' (i.e. codons) that included two sets of AAA. Weiner and Maizels emphasize that there are two equivalent tRNA-binding sites (Fig. 9.4). After the first peptide-bond

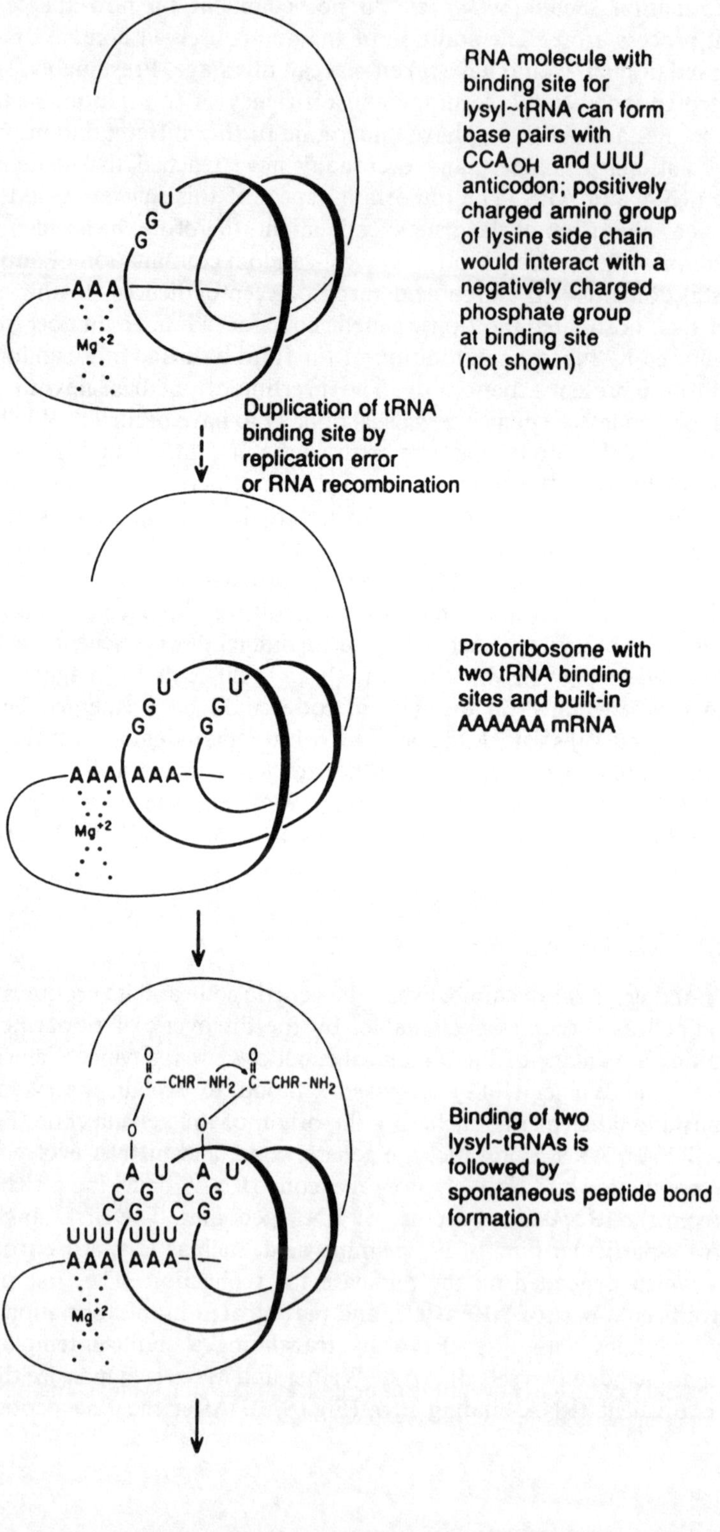
RNA molecule with binding site for lysyl~tRNA can form base pairs with CCA$_{OH}$ and UUU anticodon; positively charged amino group of lysine side chain would interact with a negatively charged phosphate group at binding site (not shown)
U
G
G
AAA
Mg+2
Duplication of tRNA binding site by replication error or RNA recombination
Protoribosome with two tRNA binding sites and built-in AAAAAA mRNA
U
G
G
U′
G
G
AAA
AAA
Mg+2
O
C–CHR–NH2
O
C–CHR–NH2
O
O
A U
C G
C G
UUU
AAA
A U′
C G
C G
UUU
AAA
Mg+2
Binding of two lysyl~tRNAs is followed by spontaneous peptide bond formation

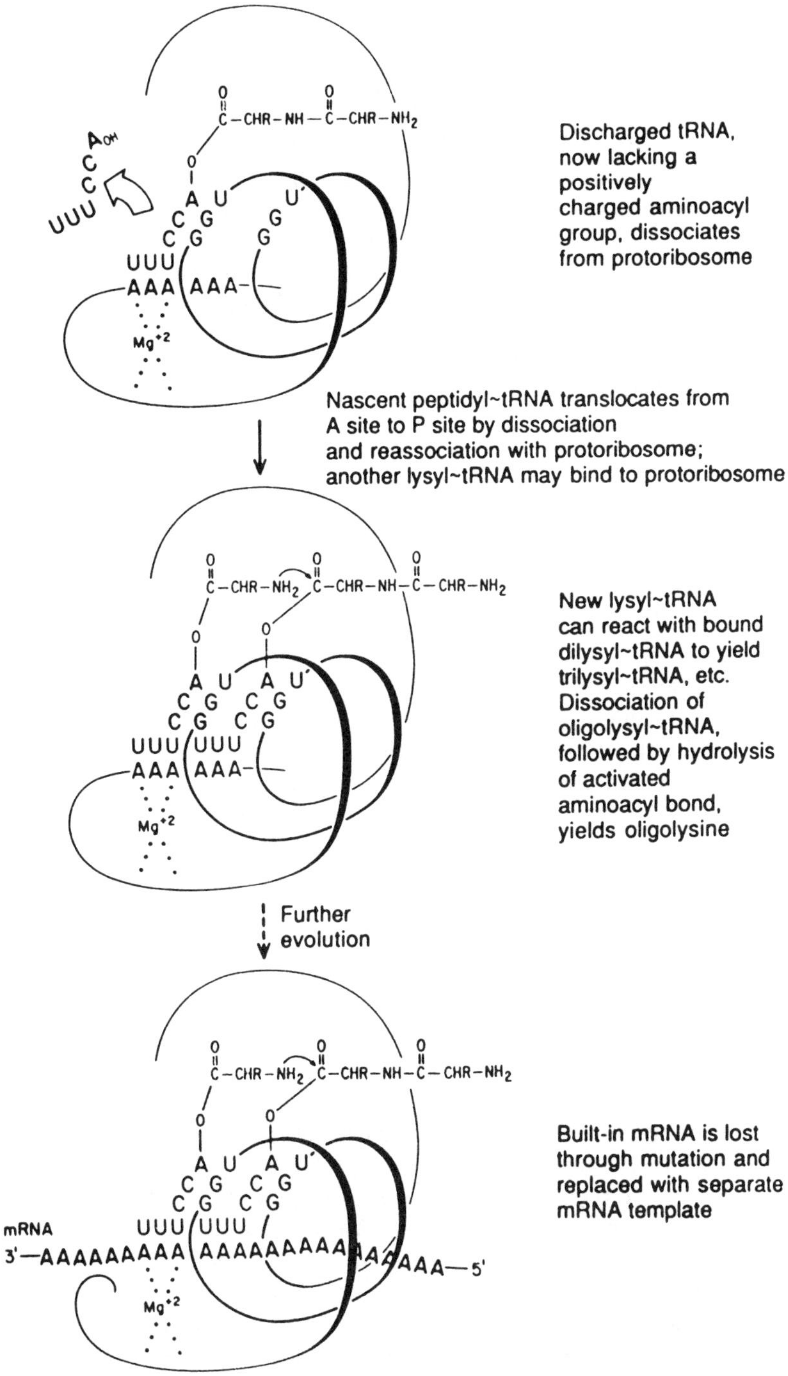

Fig. 9.4. A model for the early evolution of protein synthesizing system. The tRNAs are shown as hexanucleotides for simplicity (from Weiner, 1987).

formation, 'the discharged tRNA might dissociate and be replaced by another charged tRNA before the newly made dipeptidyl-tRNA dissociated from the (proto)ribosome . . . A second round of peptide bond formation could produce a tripeptidyl-tRNA, and so forth'. Thus, this mechanism would have involved the translocation of tRNA on the built-in-codons (corresponding to the A-site and P-site of the present-day ribosome). As only homopolymers would have been synthesized by this mechanism, the fixed coding frame and unidirectional translation do not have to be considered, unlike in a commaless code. As, in this scheme, the interaction of ARS with tRNA would determine the specificity of aminoacylation, there would be no stereochemical interaction between the amino acid and the anticodon. Therefore the code would be a 'historical accident: any triplets could have encoded a particular amino acid' (Maizels and Weiner, 1987), if an anticodon was not the primary tRNA identity determinant. Weiner (1987) emphasized the importance of basic amino acids in this protobiosynthesis of homopeptides, as positively charged amino acids might bind preferentially to the negatively charged sugar–phosphate backbone of the ARS. Then, 'if a variant tRNA synthetase arose that could activate arginine better than lysine, selective pressure would ensure that primordial tRNA structure evolved into two slightly different tRNA species . . . In this way, the rudiments of the genetic code would be defined as the population of tRNA synthetase and peptide-specific protoribosomes'. Eventually, ARS, tRNA, and protoribosomes diversified by various selection pressure to make other useful peptides. Basic peptides formed on the RNA–protoribosomes might have acted as the protoribosomal proteins by interacting with the RNA. This assumption is consistent with the fact that most of the modern ribosomal proteins are basic. During evolution, replacement of the built-in-message by a separate mRNA template, conversion of RNA–ARS to proteinous ARS, RNA–ribosome to RNP (ribonucleoprotein), etc. would have proceeded until the contemporary protein-synthesizing machinery was finally established. Weiner's model is beautifully constructed by assembling numerous experimental results of modern molecular biology. Also noteworthy would be that the genetic code, according to the model, started with a single amino acid. This notion was supported by the phylogeny of tRNA (see Section 9.1.8). The RNA-origin of ARS and ribosome would solve the problem posed by Lewin (1974) and Schimmel *et al.* (1993; see Section 9.1.5) of the emergence of proteinous ARS and translation apparatus before the emergence of the genetic code.

Evidence pointing to the direct involvement of ribosomal RNA in the function of ribosome has accumulated. Noller and co-workers (Huttenhofer and Noller, 1992) demonstrated the direct interactions of 16S rRNA with the anticodon stem-loop of tRNA. The interactions are in the A-site and P-site functions of the ribosomes. The CCA terminus of tRNA, which involves the peptidyl transferase reaction, interacts with the region near the central loop of domain V of 23S rRNA, suggesting that this region of rRNA is a component of peptidyl transferase (Moazed and Noller, 1991). Using the exhaustive removal of ribosomal proteins from *Thermus aquaticus* ribosome, Noller *et al.* (1992) found that peptidyl transferase activity was resistant. Watanabe and co-workers (Nitta *et al.*, 1994; K. Watanabe, personal communication) found that syntheses of poly(U)-

dependent polyphenylalanine, poly(A)-dependent polylysine, and alternative UC copolymer-dependent oligo Ser-Leu on *E. coli* ribosomes can each proceed in the presence of a high concentration of pyridine. Surprisingly, these reactions do not require GTP, ATP, and soluble protein factors, but the template, ribosomes, and amino-acid-charged tRNA are essential for these condensation reactions.

The mechanism of the peptidyl transferase reaction is typical nucleophilic substitution. A carbon in the ester bond between the peptide and the 3′adenosine of the peptidyl-tRNA undergoes nucleophilic attack by the lone pair of nitrogen of the amino acid moiety of aminoacyl-tRNA in the A-site of the ribosome. Pyridine should accelerate this reaction, because tertiary amines, including pyridine, are known to be useful catalysts in the nucleophilic substitution reaction (Wagner and Zoo, 1953). Dehydration condensation reactions, such as the formation of peptides and phosphodiester bonds, require a non-aqueous environment, which may be provided by the tertiary amines. Watanable *et al.* (personal communication) further demonstrated that pyridine can be substituted by adenine (or adenosine) and cytosine (or cytidine) that have the tertiary amine moieties within their structures and that ribosomes deprived of most ribosomal proteins may substitute the intact ribosomes. These findings suggest that the catalytic functions of nucleosides represent the most important reactions that might have occurred on the protoribosome.

All of these results support the idea that the protoribosome may be the ancient form of ribosomal RNA. However, how the active sites specific for amino acid in each ARS were created from the corresponding RNA-ARS is still unknown, although Weiner (1987) noted that 'there are many different ways of converting an RNA-ARS to a protein ARS'. Weiner illustrates one such possibility. First, the nucleotide cofactor was base-paired to the RNA enzyme through its adenosyl group. After evolution of a working translation system, catalysis by the original RNA enzyme–cofactor complex was enhanced by interaction with small polypeptides. Along with an increase in the reliability of the translation system, larger polypeptides—ultimately protein enzymes—took over all the RNA functions except the critical role of the cofactor in catalysis (see Fig. 28-15 in Weiner, 1987). It is important to find out if any polyribonucleotide, either in a free form or as a built-in sequence in tRNA (or minihelix), has specific aminoacyl-tRNA synthetase activity (see Illangasekare *et al.*, 1995 for a new development). Perhaps built-in mRNA was lost from protoribosome through mutations, and such ribosome underwent structural changes to accept separate mRNA template. The mechanism whereby the built-in-message is replaced by mRNA would be analogous to the replacement of the internal guide sequence with an external one in transformation of the *Tetrahymena* intron to a template-dependent RNA replicase. In this way, peptide-specific protoribosome would have evolved into the non-specific ribosome seen today.

9.1.7 *Some other theories*

A model of the ribozyme origin of the genetic code was proposed by Szathmáry (1993), which he calls 'the coding co-enzyme hypothesis'. This model is apparently based in large parts on the Weiner's model with incorporation of various theories. Like Weiner, Szathmáry postulates that the primordial aminoacyl tRNA synthetase was a type of ribozyme containing the present-day self-splicing tRNA intron and the 3′ half of tRNA. Within the ribozyme there is the 'internal guide sequence', which is complementary to the 'anticodon' (and therefore to the corresponding codon sequence, for example, GUN Val). The anticodon itself is the primordial adaptor. Adenosine is added to the 3′ end of the ribozyme synthetase and the anticodon interacts by base-pairing with the internal guide sequence ('codon') just like codon–anticodon pairing, and is followed by specific interaction of the amino acid (e.g. Val) with the anticodon (NAC for Val) by means of 'lock-and-key' mechanism. Activation of the amino acid then takes place so that the adaptor is esterified. Szathmáry states that 'the transition from anticodon to tRNA adaptors was catalyzed by the fortuitous self-splicing activity of the synthetase ribozymes. Group I introns are descendants of these ribozymes'. He further speculates that the primordial genetic code table included a few amino acids with different chemical properties. The expansion of the genetic code proceeded by ambiguity reduction through the codon capture mechanism and/or the Wong's pathway (see Section 9.1.9), and so on. Szathmáry's hypothesis does not say how ribozyme synthetase was switched to the proteinous ARS during evolution, and obviously contradicts the view that the genetic code is a 'historical accident' in which there is no direct involvement of anticodons in the specific aminoacylation reactions.

The stereochemical interaction between codons and amino acids in the origin of the genetic code was revived by Yarus (1991). Yarus's proposal is based on the observation that Arg binds stereoselectively with the guanosine site in the catalytic centre of the *Tetrahymena* rRNA intron, the representative of the group I RNAs. The binding site consists only of AGR or CGR in the known 92 group I RNA sequences. Yarus suggests that the genetic code for Arg could be originated by this sequence-specific, Arg-specific RNA complex (protoribosome) related to the group I RNAs. As it is not likely that all amino acids interact specifically with their codons, Yarus introduces anticodons as an addition to the repertoire of specifically bound amino acid chains: Codons and anticodons exist adjacent to each other in a helical groove, on complementary strands of RNAs. Anticodons with affinity for amino acid side-chains combine with the corresponding amino acids, which may be utilized for their polymerization ('codonic period', according to Yarus). This addition is particularly necessary when there is little amino acid–codon interaction. This, in turn, seems to mean that direct amino-acid–codon interaction is not always required. One strand with anticodons will become the progenitor of the tRNAs by fragmentation of the strand and attachment of amino acids (by a fragment of RNA = proto-ARS) to the termini of the anticodons. Another strand with codons will become the progenitor of mRNA, thus resulting in a clear distinction between mRNA and complementary adaptors ('anticodonic period'). Further development of the coding system led it to the recent 'peptidic period'. According to Yarus, Arg was one of the first amino acids to enter the genetic code through direct interaction of universal Arg codons. However, codons CGN and AGR—the universal Arg codons—are the most changeable in their assignment. CGN and AGA sometimes become unassigned, and AGR is sometimes changed to codons for Ser, Gly, or stop. Thus there is no guarantee that these codons originated as Arg codons. Affinity of anticodons to amino acid side-chains, which is postulated to have occurred in the codonic period, is not convincing, although Wong (1988) adopted this explanation. Experimental evidence is obviously needed to prove or disprove such an affinity.

9.1.8 *Evolution of tRNA and amino acid tRNA synthetase*

The secondary structures of all the known tRNAs, except for some non-plant mitochondrial tRNAs (see Fig. 1.1), have the following characteristics in common:

1. Four helical regions containing 7, 4, 5, and 5 base pairs, respectively, in which all but two pairs are of variable composition.
2. Four loops of unpaired nucleotides, of which the third loop varies in size and may contain up to 17 nucleotides.
3. A total of 23 invariant or semi-invariant nucleotides. The tertiary structure is constant.

Given these properties, it seems that tRNAs are descended from a single ancestral molecule, and diversified by gene duplications and mutations. Fitch and Upper (1987) deduced the 'urancestral tRNA' from the 300 tRNA sequences available, and Eigen *et al.* (1989) placed the earliest tRNA at not more than 3.8 ± 0.5 billion years ago. Their conclusion was that 'the genetic code is not older than, but almost as old as, our planet.'

Phylogenetic analyses (Nagel and Doolittle, 1991) show that the present ARSs were derived from two independent ancestors (Fig. 9.5).

The tRNA and ARS phylogenies show their diversification profiles but do not necessarily represent the evolution of the anticodon-dependent genetic code, because a limited number of amino acids could have been aminoacylated on the acceptor site of RNA before joining of two RNA domains, i.e. the genetic code-dependent protein synthesis could have come later than the RNA code-dependent aminoacylation, according to the RNA code hypothesis.

Evolution in sequences of tRNAs has proceeded so far that the average divergence between pairs of tRNAs for different amino acids has reached an equilibrium between forward and backward mutation of variable sites (Holmquist *et al.*, 1973). In addition, a relatively short chain length of tRNA makes it difficult to construct a reasonable phylogenetic tree. When the tRNA phylogeny is established, it will provide a deep insight into the evolution of the genetic code.

9.1.9 *Evolution of the ancient genetic code*

It is possible to speculate from the tRNA phylogeny, that the genetic code might have started with a single amino acid. Independent of the phylogenetic studies, Weiner (1987) has assumed a single amino acid origin, as discussed above.

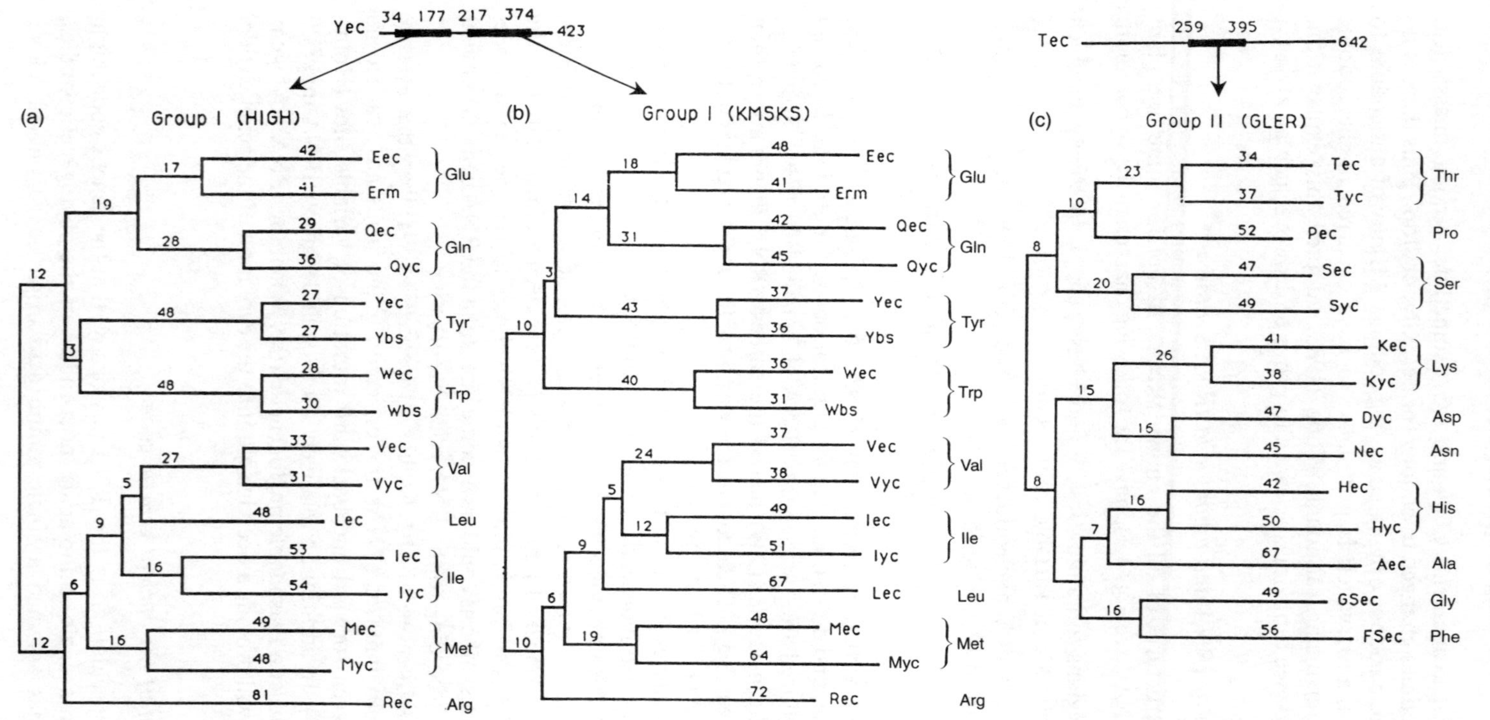

Fig. 9.5. Phylogenetic trees of aminoacyl-tRNA synthetases. Trees for two sequence segments, HIGI and KMSKS, were constructed separately for group I enzymes. One segment, GLER, was used for the tree of group II enzymes. The segments for Yec and Tec are shown above the trees. Homologous segments of various ARSs were used for construction of the trees. Distance scores were derived from the parsimony-based PAPA program. Abbreviations: the single-letter symbol for the amino acid along with ec for *Escherichia coli*, bs for *Bacillus stearothermophilus*, rm for *Rhizobium meliloti*, and yc (yeast) for *Saccharomyces cerevisiae*; e.g. *E. coli* valyl-tRNA synthetase is Vec, etc. (from Nagel and Doolittle (1991), with modifications).

However, there would have been a considerable number of amino acids in existence before the genetic code developed. Bearing this in mind, it has been suggested that the coding system arose with a limited number of amino acids, and that others were added until the total of 20 was reached, presumably after the development of ribosome and proteinous ARSs.

Although many introductions of 'new' amino acids into proteins would be disruptive for organisms, it is also true that an increase in amino acid content must be allowed for the introduction of new amino acids at some stages. The code finally would have reached to include the present-day 20 amino acids. This would imply that the code was frozen in one respect: the same 20 amino acids are in all codes, including the mitochondrial code. No other amino acids could enter the code, with the exception of Sec, for which, however, only an optional use of UGA is allowed.

Phylogenetic analyses of ARSs indicate that some ARS 'pairs', typically for Glu/Gln and Asp/Asn, cluster rather closely. Nagel and Doolittle (1991) suggest that some new amino acids were added relatively recently to protein synthesis.

The freezing of the code for the 20 amino acids would have occurred when the present ARSs and tRNAs for the 20 amino acids became available. At this stage, the code determined the amino acid sequences of so many highly evolved protein molecules that any addition of new amino acids would not be tolerable. The changes in the amino acid assignments of codons were not possible unless they occurred by a neutral or near-neutral process without altering the amino acid sequence of proteins (see Section 6.4).

Wong (1975, 1976, 1981, 1988) proposed the co-evolution theory of the code with incorporation of stereochemical mechanisms. He postulated that first only six to ten 'phase 1' amino acids (Glu, Asp, Val, Ser, Ala, and Gly, and probably Pro, Thr, Leu, and Ile) were assigned by 61 codons, the remaining three codons being for stop signals. Thus, more than four synonymous codons existed for each amino acid in this primitive code. The code then expanded by the addition of other 'phase 2' amino acids; so Tyr was formed by the known biosynthetic pathways from Phe; Cys and Trp from Ser; Lys, Thr, and Asn from Asp; Arg, Pro, and Gln from Glu; His from Gln; Met and Ile from Thr; and Leu from Val (Table 9.3). These new amino acids acquired the codons contiguous with those of their precursor amino acids in many cases. For example, CCN (presently Pro), CAN (His and Gln), CGN (Arg), AGR (Arg), and GAR (Glu) were all the codons for Glu, and the phase 2 amino acids (Pro, Gln, Arg) are derived from Glu through biosynthetic pathways and receive codons from those of Glu; His was later formed from Gln. Wong emphasized the importance of relative hydrophobicity in selective association between amino acids and anticodons; a highly ranked amino acid tends to be associated with a highly ranked anticodon (it is not always the case, however). He noted that:

> Where product (amino acid) resembles precursor structurally (by hydrophobicity), competition between them for attachment to the tRNAs of the precursor sufficed to transfer codons from precursor to product. Where resemblance is weak, a pretranslational

Table 9.3 Classification of phase 1 and phase 2 amino acids

Amino acid	Membership in code	Synthesis by electric discharge	Presence on meteorites	Biosynthetically produced from
Gly	early	+ + +	+ + +	–
Ala	early	+ + +	+ +	–
Ser	early	+	–	–
Asp	early	+	+	–
Glu	early	+	+ +	–
Val	early	+	+ +	–
Pro	probably early	+	+ +	Glu
Thr	probably early	+	–	Asp
Leu	probably early	+	+*	Val
Ile	probably early	+	+*	Thr
Phe	probably late	–	–	–
Tyr	probably late	–	–	Phe
Cys	probably late	–	–	Ser
Lys	late	–	–	Asp
Arg	late	–	–	Glu
His	late	–	–	Gln
Met	late	–	–	Thr
Trp	late	–	–	Ser
Asn	late	–	–	Asp
Gln	late	–	–	Glu

* Stated as (–) by Wong. See Table 9.1.
From Wong (1988) with modifications.

modification of the precursor tRNA to form product tRNA would achieve the desired result [e.g. Glu-$tRNA^{Glu}_{UUC}$→Glu-$tRNA^{Gln}_{UUG}$→Gln-$tRNA^{Gln}_{UUG}$; changes of a precursor Glu (the least hydrophobic) to a product Gln (hydrophobic at the third rank) while linked to the tRNA].

Unlike Wong, Weber and Miller (1981) did not offer a biosynthetic addition of various amino acids, such as Lys, Arg, Trp, and Met, which could not be formed by prebiotic synthesis. They suggested that these amino acids could have been produced by various prebiotic syntheses.

The observation by Wong that related codons (convertible by single base change) generally, but not always, specify functionally related amino acids (see Epstein, 1966) suggests that early proteins composed of a limited number of amino acids could be transformed, by a near-neutral process resulting in the addition of new amino acids, to more accurate proteins without much disruption of the pre-existing proteins. The fact that the phylogenetically related ARSs recognize chemically similar amino acids (Nagel and Doolittle, 1991) is consistent with this view.

Another view, which is not mutually exclusive, is that the early protein synthesizing system was crude, so that aminoacylation of early tRNAs was more or less ambiguous. Nagel and Doolittle (1991) suggest that there would have been some ambiguity in the recognition of amino acids by an ARS before radiation of ARS. The primordial ARS initially charged a tRNA or a group of tRNAs with two or more amino acids of similar chemical properties, so that the changes in protein sequence could be tolerated. Such ambiguity was reduced by refinement of specificity during the subsequent evolution of the genetic code.

Some ambiguity would certainly have existed in the ancient genetic code. For example, Gln is formed by the amidation of glutamyl (Glu)-tRNA^{Gln}, which is charged with Glu by glutamyl-tRNA synthetase in *Bacillus subtilis*, *B. megaterium*, *Lactobacillus acidophilis*, and perhaps *Mycoplasma capricolum*. These bacteria do not seem to possess glutaminyl (Gln)-tRNA synthetase (Wilcox, 1969). This suggests either that glutaminyl-tRNA synthetase disappeared during evolution, or that all these Gram-negative bacteria had separated from the ancestor before glutaminyl-tRNA synthetase evolved from glutamyl-tRNA synthetase, and that they used the 'primitive acylation system' for Gln without the emergence of glutaminyl-tRNA synthetase. In which case, why is the Glu attached to tRNA^{Glu} not amidated? Presumably, in the early protein synthesizing system, both tRNA^{Glu} and tRNA^{Gln} accepted Glu, and some of these were randomly amidated, so that GAR and CAR codon sites specified Glu or Gln without distinction. Such a non-specific amidotransferase evolved to react only with Glu-tRNA^{Gln}. It is possible that amidation of Glu attached to tRNA^{Gln} in *B. subtilis* and others is a remnant of the ancient system.

Fitch and Upper (1987) suggested a fully ambiguous ancestral code with 'no particular preference of any amino acid for any particular codon'. The ambiguity would have been reduced until the present form of the genetic code was reached. Fitch and Upper then deduced the urancestral tRNA sequence responsible for a fully ambiguous code, although it is not very likely that such a fully ambiguous code existed. Fitch and Upper suggest that the anticodon of this tRNA must have paired with all the codons, and that the tRNA must have been charged with any amino acid. However, neither such an ambiguity in anticodon–codon pairings, nor the entirely non-specific aminoacylation of a tRNA has been found.

Jukes (1966, 1981) proposed that the code in some early stage (archetypal code) consisted of 16 family boxes, one box being for four stop codons, and the other 15 being each occupied by a single amino acid with its four codons, which were translated by a single tRNA with a UNN anticodon (Table 9.4). Subsequently, this code expanded by dividing some of the family boxes into 2-codon sets, resulting in the early code (see below) for 20 amino acids. For this expansion, the single tRNA genes had to duplicate, and, under GC-pressure, one of the duplicates underwent a mutational change from UNN to GNN in the anticodon. For example, if Gln was in a family box (CAN) with anticodon UUG, the anticodon would duplicate to UUG and GUG, and the tRNA with anticodon GUG would be captured by His. The other anticodon, UUG, became modified to *UUG so that it would not continue to pair with all four CAN codons, but only with CAA and

Table 9.4 Anticodons of archetypal code

Anticodon	Amino acid	Anticodon	Amino acid
UAA	Phe or Leu	GUA	Tyr
UAG	Leu	UUG	His or Gln
UAU	Ile or Met	UUU	Asn or Lys
UAC	Val	UUC	Asp or Glu
UGA	Ser	UCA	Cys or Trp
UGG	Pro	UCG	Arg
UGU	Thr	UCU	Arg or Ser
UCG	Ala	UCC	Gly

Each anticodon pairs with four codons for a single amino acid, except for GUA, which pairs only with AUU and AUC.
From Jukes (1983*a*).

CAG. Naturally, this scheme does not necessarily mean that the archetypal code was the most ancient one, but it is reasonable to assume that this code preceded the code for 20 amino acids. This scheme allows some changes in the amino acid sequences of pre-existing proteins upon addition of new amino acids. However, an organism at this stage of development would be quite simple and primitive. If the organism could survive this change, acquisition of the new amino acids in the genetic code might provide for an evolutionary advantage (Jukes, 1983*b*).

9.1.10 *Summary*

It is still premature to draw any plausible picture of the origin and early evolution of the genetic code from the widespread and often controversial theories discussed above. A simple summary is given below:

1. A good number of amino acid species were available in the prebiotic period before emergence of the genetic code.
2. tRNAs are probably descended from a single ancestral molecule, while ARS are derived from two independent ancestors.
3. ARSs and ribosomes probably originated from ribozyme-like RNA molecules before they evolved into the present ARSs and ribosomes.
4. The specific aminoacylation of tRNAs is independent from the anticodons in many cases. Thus, the stereochemical theory of evolution of the genetic code seems to be no longer tenable, because the theory implies that the amino acid assignments of codons are more or less predetermined by specific and direct interaction of amino acids with codons and/or anticodons; there is no con-

vincing evidence for this. Instead, the amino acid assignments of codons in the present genetic code would have been brought about by accident.

5. Whether the genetic code started with a single amino acid or with a limited number of amino acids is not clear.
6. Fully ambiguous ancestral genetic code is not likely, but there might have been some ambiguities in aminoacylation of tRNAs in the ancient genetic code.
7. The addition of 'new' amino acid took place during evolution until 20 amino acids were frozen into the code. This occurred in a single progenote that displaced all contemporaries with other codes. The prototype of the universal genetic code (the early genetic code; see Section 9.2) was thus established.

9.2 The early genetic code

The code that existed shortly before the universal genetic code was reached may be reconstructed from the knowledge now available.

It is reasonable to assume that in the early genetic code translation of codons to 20 amino acids was performed more simply than in the present, highly evolved code, using minimum or near-minimum number of tRNA species. The number of anticodons as recognized by isoacceptors has increased during evolution with increasing complexity of the genome, so that the fidelity, efficiency, and other regulation by codon–anticodon pairing have improved in various ways.

One of the most characteristic features of the universal genetic code is its definite pattern. Each pair of codons ending with a pyrimidine specifies a single amino acid; this regularity exists for all pairs of codons ending with a purine, with the exception of the pairs AUA (Ile)/ AUG (Met) and UGA (stop)/ UGG (Trp).

Jukes (1983*a*) proposed, on the basis of the most simple anticodon composition of vertebrate mitochondria among various present-day coding systems, that the mitochondrial genetic code was brought about by retrogression from the universal genetic code to the early genetic code through genomic economization, which often accompanies AT-pressure. It is possible, then, that the early genetic code evolved to the present code by nearly the reverse process of the mitochondrial genetic code evolution (see Osawa *et al.*, 1988, 1992). The retrogression process of the mitochondrial genetic code would have involved two main events. The first event is the reassignment of AUA codon from Ile to Met, and UGA from stop to Trp, as explained in Sections 6.3 to 6.6. This suggests that AUA and AUG were for Met and UGA and UGG were for Trp in the early genetic code (Table 9.5). Exactly the same table for the genetic code immediately preceding the universal genetic code was proposed by Hasegawa and Miyata (1980), by taking the mitochondrial genetic code as a fossil evidence of the earlier code. Both vertebrate mitochondrial genetic code and the early genetic code removes the exceptions mentioned above. The second event in the retrogression process of the mitochondrial genetic code is the reduction of the number of anticodon

Table 9.5 The early genetic code

Amino acid	Codon	Anticodon	Amino acid	Codon	Anticodon	Amino acid	Codon	Anticodon	Amino acid	Codon	Anticodon
Phe	(UUU)	GAA	Ser	(UCU)	UGA	Tyr	(UAU)	GUA	Cys	(UGU)	GCA
Phe	(UUC)		Ser	(UCC)		Tyr	(UAC)		Cys	(UGC)	
Leu	(UUA)	*UAA	Ser	(UCA)		Stop	(UAA)	—	Trp	(UGA)	*UCA
Leu	(UUG)		Ser	(UCG)		Stop	(UAG)	—	Trp	(UGG)	
Leu	(CUU)	UAG	Pro	(CCU)	UGG	His	(CAU)	GUG	Arg	(CGU)	UCG
Leu	(CUC)		Pro	(CCC)		His	(CAC)		Arg	(CGC)	
Leu	(CUA)		Pro	(CCA)		Gln	(CAA)	*UUG	Arg	(CGA)	
Leu	(CUG)		Pro	(CCG)		Gln	(CAG)		Arg	(CGG)	
Ile	(AUU)	GAU	Thr	(ACU)	UGU	Asn	(AAU)	GUU	Ser	(AGU)	GCU
Ile	(AUC)		Thr	(ACC)		Asn	(AAC)		Ser	(AGC)	
Met	(AUA)	*UAU	Thr	(ACA)		Lys	(AAA)	*UUU	Arg	(AGA)	*UCU
Met	(AUG)		Thr	(ACG)		Lys	(AAG)		Arg	(AGG)	
Val	(GUU)	UAC	Ala	(GCU)	UGC	Asp	(GAU)	GUC	Gly	(GGU)	UCC
Val	(GUC)		Ala	(GCC)		Asp	(GAC)		Gly	(GGC)	
Val	(GUA)		Ala	(GCA)		Glu	(GAA)	*UUC	Gly	(GGA)	
Val	(GUG)		Ala	(GCG)		Glu	(GAG)		Gly	(GGG)	

species during mitochondrial evolution. The U-modification enzymes for $^{+}$UNN anticodons in family boxes were removed, so that the resultant UNN anticodons (U unmodified) translated all four codons. The GNN anticodons that had translated NNY codons became unnecessary and were eliminated. *UNN anticodons for NNY codons and GNN for NNR codons were retained in 2-codon sets because NNY and NNR code for different amino acids. All redundant CNN anticodons, except Met CAU, were also removed because of AT-pressure. This resulted in 22 anticodon species. The minimum number of anticodons for translating 62 codons (UAR is for stop) corresponding to 20 amino acids is 23, but in the vertebrate mitochondrial genetic code AGR are stop codons instead of Arg codons. The code table of some mitochondrial genetic codes, e.g. in filamentous fungi, is the same as that of the postulated early genetic code (see Tables 3.1 and 3.4). In *Mycoplasma capricolum*, UGA is a Trp codon and its code table differs from the early genetic code in using AUA for Ile and in CGG being an unassigned codon. It has similar anticodon compositions, although slightly more complex, than the early genetic code. The mitochondrial and *Mycoplasma* genetic codes use a single UNN anticodon per family box in all or most cases. In chloroplasts, the codon table is the same as the universal code, but some family boxes use a single UNN anticodon. It is notable that most of the CNN anticodons are removed in all these coding systems. These facts suggest that these organisms or organelles are generally under AT-pressure, and are in the course of retrogression in various degrees. With respect to the similarity of the anticodon compositions of the early genetic code with that of *Mycoplasma* and other organelles, it may be assumed that the progenote genome, in which the early genetic code was used, was also under AT-pressure.

It is not clear whether all 64 codons were used in the early genetic code, because Leu UUR/CUN, Arg CGN/AGR, and Ser UCN/AGY are redundant, and the removal of one of them does not affect the number of amino acids used for protein synthesis. Some, e.g. CUN (Leu) and CGN (Arg), could have been unassigned codons, i.e. non-existent codons (because of their possible removal by AT-pressure), which were later assigned to the respective amino acids. In any case, the anticodon composition of the early genetic code would be similar, if not identical, to the vertebrate mitochondrial genetic code. In short, in the early genetic code, four codons in family boxes could be translated by a single anticodon UNN, and codons NNY and NNR in the 2-codon sets could be read by GNN and *UNN anticodons, respectively. The number of anticodons was 23 (or less if some codons were not used (unassigned codons)).

9.3 From the early genetic code to the universal genetic code

9.3.1 *Problems to be solved*

The main problems to be solved in the evolutionary route from the early genetic code to the universal genetic code are: (1) reduction of the number of codons from two to one for Trp (UGA + UGG to UGG) and for Met (AUA + AUG to

AUG); and (2) reassignment of AUA to an Ile codon and of UGA to stop codon (Osawa *et al.*, 1988, 1992).

The changes in the reassignment of codon AUA from Ile to Met in the mitochondrial genetic code and of codon UGA from stop to Trp both in mitochondrial and *Mycoplasma* codes most probably proceeded with high AT-pressure as the driving force. The reverse process—AUA Met to Ile and UGA Trp to stop—could have occurred under GC-pressure during the evolution of the early genetic code to the universal genetic code. The GC-pressure is therefore assumed to be important in the evolution of the genetic code after the early genetic code was reached.

9.3.2 *Reduction in the numbers of Trp and Met codons*

With increasing GC-pressure, the Trp tRNA anticodon *UCA, which pairs primarily with codon UGA and poorly with UGG, became unable to accommodate the increasing number of Trp UGG codons. The gene for tRNA with anticodon *UCA adaptively duplicated and one of the duplicates, under GC-pressure, mutated to CCA, which translates only codon UGG. Finally, UGA codons were converted to UGG under GC-pressure and positive selection pressure by anticodon CCA. Anticodon UCA was no longer needed and was discarded, leaving codon UGA unassigned. Any back-mutation from UGG to UGA in reading frames should be deleterious or lethal, because anticodon CCA can translate only codon UGG as Trp. This explains why Trp has only one codon.

Similarly, under GC-pressure, AUA codons were all converted to AUG, which at this stage would be the sole codon for Met. Met tRNA anticodon *UAU was changed by mutation to CAU, pairing with Met codon AUG. AUA thus became an unassigned codon.

The above processes are analogous to the production of unassigned codons AGA and AUA which are actually observed in *Micrococcus luteus* with a genome high in G and C (Kano *et al.*, 1993).

9.3.3 *Introduction of the UGA stop codon*

After replacement of anticodon UCA by CCA, some UAA stop codons mutated to UGA under GC-pressure, which for the first time became a stop codon. At this stage, RF-2 would have emerged so as to recognize both UAA and UGA as stop codons. Thus UGA obtained its chain termination function, because its location was identical to the ancestral UAA stop. There was no UCA anticodon for Trp to pair with UGA, because it had been discarded. Thus, stop codon UGA would have originated from UAA stop as an additional stop codon, unrelated to the primitive Trp codon UGA. There are examples of conversion of UAA to UGA in many protein genes. UGA is the predominant stop codon in high G + C bacteria such as *Micrococcus luteus*, replacing UAA stop codon (Ohama *et al.*, 1989). This suggests conversion of TAA to TGA by GC-pressure.

9.3.4 *Reassignment of codon AUA for Ile*

The next question is the reassignment of codon AUA to Ile. First, the $tRNA^{Met}_{CAU}$ duplicated. In one of the duplicate tRNAs, C in the first position of this anticodon became modified to LAU upon emergence of lysidine-forming enzyme. This change resulted in the tRNA accepting Ile instead of Met and pairing with A instead of G (Muramatsu *et al.*, 1988*a*,*b*). Now, AUA codons could be produced from AUY Ile codons by mutations and were captured by the new $tRNA^{Ile}_{LAU}$.

Propagation of the new AUA Ile codons would have taken place during the subsequent evolution to various organismic lines, especially those with genomes high in A and T. For example, in *E. coli* (G + C = 50 per cent), AUA is a rare codon, and AUC is used abundantly. In contrast, considerable utilization of AUA is seen in *Mycoplasma* spp. (G + C = 25 per cent) or in *Bacillus subtilis* (G + C = 43 per cent).

The genes for $tRNA^{Met}_{CAU}$ and $tNRA^{Ile}_{LAU}$ are adjacent on the chromosomes of *Bacillus subtilis*, *Mycoplasma capricolum*, and *Spiroplasma* sp. (Muto *et al.*, 1990). This fact favours the ancient duplication of the gene for $tRNA^{Met}_{CAU}$ and the subsequent conversion of one of them to $tRNA^{Ile}_{LAU}$.

9.3.5 *Evolutionary diversification of anticodons*

The preceding changes must have occurred in the progenote before diversification, because the uses of codon AUA for Ile and UGA as a stop codon are common to all organismic lines, except for the changed codons in mitochondria and *Mycoplasma*. The above changes completed the universal genetic code, presumably some 20 million years ago.

The anticodon list at this stage would have contained GNN for NNY codons in a family box. This view is supported by the fact that both eubacteria and archaebacteria (metabacteria) use GNN anticodons in family boxes, while in eukaryotes GNN is replaced by INN, except for Gly GGY. As archaebacteria are closer to eukaryotes than to eubacteria (Hori and Osawa, 1979, 1987; Iwabe *et al.*, 1989), one must assume that the common ancestor (the progenote) to eubacteria and archaebacteria used GNN in family boxes, unless the GNNs developed independently in the two bacterial lines. Presumably, anticodon GCC for GGN Gly codons in eukaryotes is a remnant of the ancestral anticodon GCC.

The emergence of GNN anticodons could be accomplished by duplication of the genes for $tRNA_{UNN}$ and subsequent mutation of one duplicate to GNN under GC-pressure. This improved the translation efficiency of NNY codons, because reading of NNC codons by UNN anticodons is often not efficient.

The anticodon list would also have contained some CNN anticodons, because GC-pressure was needed for evolution of the early genetic code to the universal genetic code.

After establishment of the universal genetic code in the progenote, diversification took place into three main groups—eubacteria, archaebacteria, and eukaryotes. The code evolved along with these groups, resulting in the differences of their anticodon compositions. In short, the 'typical' eubacterial genetic code evolved in such a way that anticodons ICG/CCG for the Arg family box replaced GCG/UCG; no such replacement took place in the archaebacterial line. The main differences of the eukaryotic genetic code from the two bacterial genetic codes are that eukaryotes use INN anticodons for all family boxes and for AUY/AUA Ile codons, except for Gly GGY, and use CNN anticodons for nearly all NNG codons. A possible explanation of the latter was described in Section 2.2 as a low ability of anticodon *UNN to pair with NNG codon on the eukaryotic ribosomes.

The evolution of the genetic code has continued within each of the three main groups, quite often affected by directional mutation pressure. For example, in the case of eubacterial and archaebacterial codes, evolution has led to differences in the content of CNN anticodons in different bacterial families, and higher G + C content of DNA by GC-pressure is accompanied by larger numbers of CNN anticodons.

The mitochondrial and chloroplast genetic codes have evolved through retrogression from the eubacterial genetic code, as described above and in Section 3.3. There are 31 anticodons in the chloroplast code (see Table 3.2), compared with 22 in the vertebrate mitochondrial code (see Table 3.4). One of these—Pro anticodon GGG—is present in a pseudogene, so the only anticodon for Pro is presumably UGG. Most of the CNN anticodons in the ancestral eubacterial code have disappeared during evolution of the chloroplast code, presumably under AT-pressure and genomic economization. Four-way wobble is used in three family boxes in the chloroplast code, which has apparently evolved from an ancestral eubacterial code along a pathway similar to that of the mitochondrial code, although the evolutionary process has not proceeded as far.

9.3.6 *Origin of inosine-containing anticodons*

Alone of the amino acids, Arg has an ICG anticodon in eubacteria. How this developed, and why eukaryotes have seven other INN anticodons in addition to ICG, are significant. At about the time of establishment of the universal genetic code, Arg had anticodon GCG pairing with codons CGU and CGC, and anticodon UCG pairing with codons CGA and CGG (see p. 157 and Fig. 9.6). Under continuing GC-pressure, CGA codons disappeared by mutation to CGG and CGC, and became unassigned along with mutation of anticodon UCG to CCG. Later, upon relaxation of GC-pressure, anticodon ICG (ACG on DNA) appeared by mutation from GCG. Anticodon ICG paired with Arg codons CGU, CGC, and CGA. The fourth Arg codon CGG continued to pair with anticodon CCG.

The eukaryotic INN anticodons were probably created in a way similar to eubacterial ICG. The principles underlying these changes are analogous to those for AAA from Lys to Asn, AUA from Met to Ile as a reversal of Ile to Met, and UAA from stop to Tyr in mitochondria, in all of which anticodon *GNN (*G, modified G) would have to pair with NNY/A codons with CNN and NNG codons. However, it is difficult to decide whether a system forming inosine from adenosine developed independently in the two groups. Archaebacteria use anti-

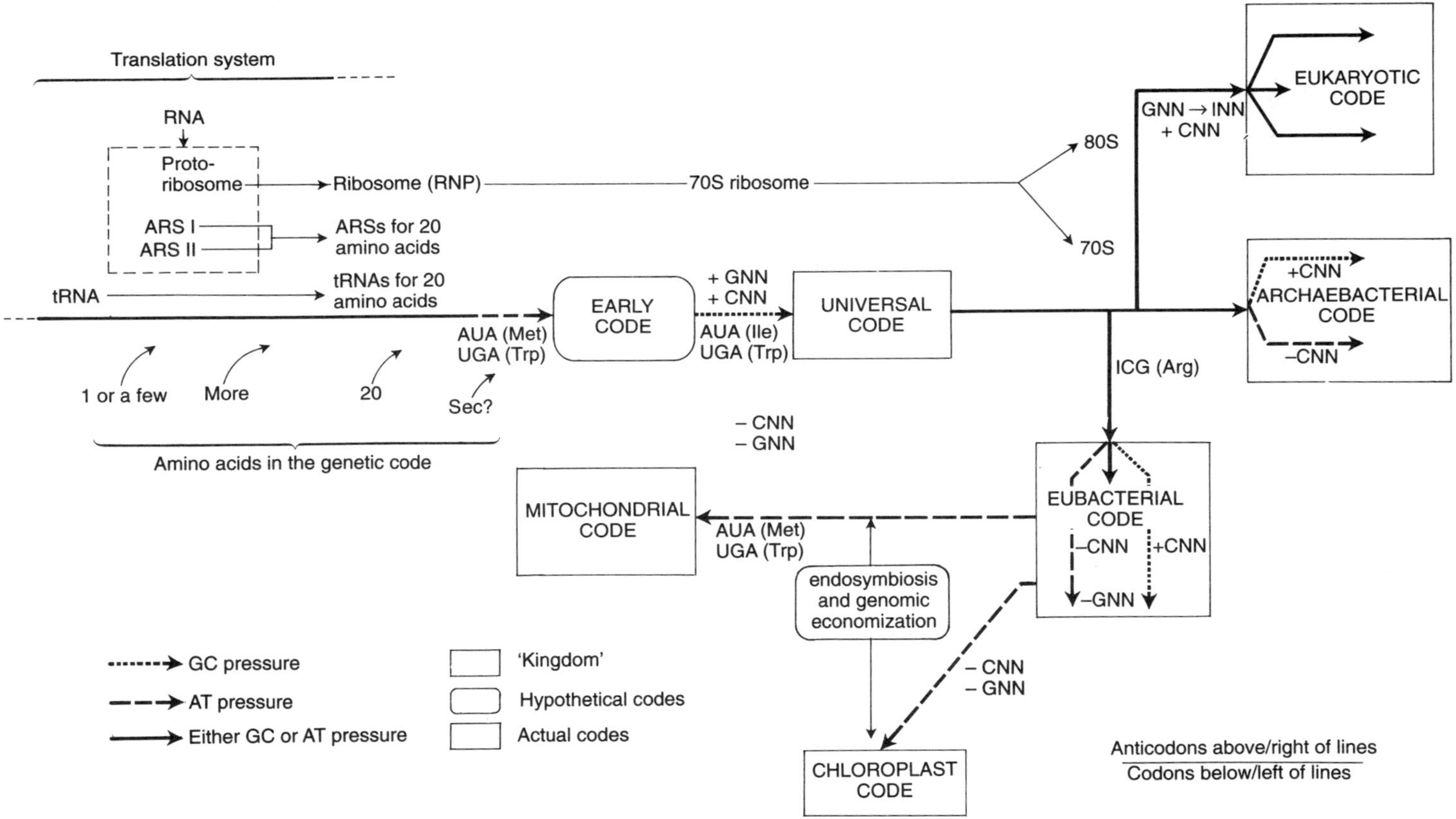

Fig. 9.6. Evolution of translation apparatus, genetic codes, showing periods of GC- and AT-pressure, and the main changes in codon assignments and tRNA anticodons. Details in evolution of tRNA anticodons and non-universal genetic codes in each 'kingdom' and organella are not indicated (see text) (based on Osawa and Jukes, 1989).

codon GCG instead of ICG. This would favour an independent origin, but it is also possible that anticodon GCG in high G+C archaebacteria such as *Halobacterium* could be formed from back-mutation of ACG under GC-pressure.

Conversion of anticodon GAU to IAU in the Ile box provided pairing for codons AUU, AUC, and AUA, so that anticodon LAU pairing with codon AUA used in the universal genetic code of the progenote became unnecessary and disappeared in eukaryotes.

The whole 'story' of the evolution of the genetic code is summarized in Fig. 9.6.

10 Amino acid composition of proteins and the genetic code

Constraints that affect the amino acid composition of proteins include the composition of the genetic code, the relation of amino acid sequence and content to function, the existence of variable and conserved regions, and the presence of GC/AT-pressure in the genome (see below). Some proteins, e.g. the histones, evolve much more slowly than others, e.g. fibrinopeptides. Near-neutral amino acid changes are found in variable regions of proteins and are more common in rapidly evolving proteins.

10.1 Amino acid composition of proteins—the role of the genetic code

The most important constraint is the composition of the genetic code. Amino acids with four codons, such as Ala, Val, and Thr, occur more frequently in

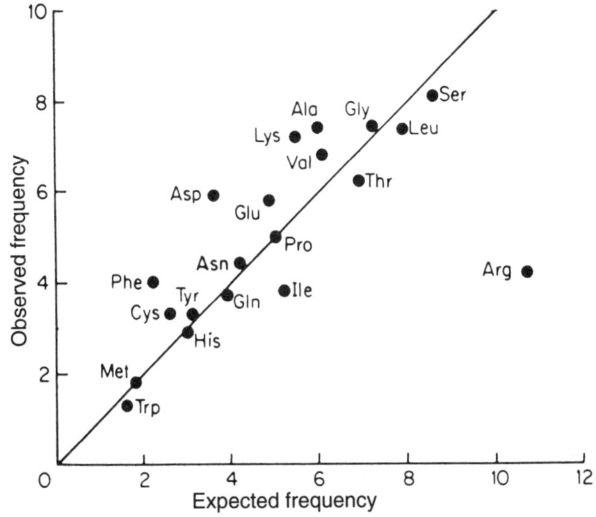

Fig. 10.1. The similarity between the observed frequencies of amino acids in 53 mammalian proteins and the frequencies predicted by the genetic code and random permutations of DNA nucleotides. The frequencies are in percentages of total amino acid content. The straight line represents an idealized equality of expectation and observation. There is good agreement, with the exception of Arg (see text) (from King and Jukes, 1969).

proteins than do Met and Trp, which have only one codon apiece. Indeed, there is fairly good agreement between amino acid frequencies in protein and those expected from their relative number of codons in the genetic code (King and Jukes, 1969; Fig. 10.1). King and Jukes state that 'in the non-Darwinian model of evolution, the amino acid composition of protein should be strongly influenced by the code, since a significant proportion of amino acids in proteins would be present as a result of random mutation and drift'. More use of Val than Trp occurs because Val has four times (twice in some codes) as many codons as Trp, and *not* because the number of Val codons depends upon a higher demand for Val by proteins. The composition of codons in the genetic code is therefore largely responsible for the amino acid content of proteins. It is misleading to think that the composition of the code evolved in order to best match the amino acid requirement of proteins, as may be argued from the stand-point of the Darwinian model.

10.2 Selection against the genetic code—functional constraint

The constraint by the genetic code is modified, or even magnified, by the fact that the average composition of proteins deviates from the proportions of codons in the genetic code. As a result, the percentages of each of the four nucleotides in the first and second codon positions are not uniform. This is shown in Table 10.3, based on 'average' or 'consensus' protein compilated from 47 eukaryotic, 17 bacterial and 4 virus proteins. This protein contains per 61 ($\times$10) residues: Ala-53 Arg-26 Asn-30 Asp-36 Cys-13 Gln-24 Glu-33 Gly-48 His-14 Ile-31 Leu-47 Lys-41 Met-11 Phe-25 Pro-25 Ser-45 Thr-37 Trp-8 Tyr-23 Val-42 (Jukes *et al.*, 1975). In making the calculations, it was assumed that all codon sets were divided equally among synonymous codons. As a result, T, C, A, and G should be approximately equally represented in third positions, although the exclusion of stop codons will make a slight difference in the percentages.

Percentages of A and G are high in the first position (see Table 10.3; columns of consensus protein) because Lys, Asp, and Glu levels are high in proportion to their representation in the code (Fig. 10.2). A is high in the second position for the same reason, and G is low because Arg is low (Fig. 10.2).

The content of the basic amino acids Arg and Lys in 68 proteins from eukaryotes, prokaryotes, and viruses was 4.3 per cent and 6.7 per cent, respectively, a total of 11 per cent compared to 9.8 per cent plus 3.3 per cent, totalling 13.1 per cent in the genetic code. The content of the acidic amino acids Asp and Glu was 5.9 per cent and 5.4 per cent, a total of 11.3 per cent, which is about twice as much as the genetic code (which contains 6.6 per cent). Thus the average electrostatic charge is roughly neutral, while in the genetic code Lys plus Arg predominate in a ratio of 2 : 1 over Asp plus Glu. However, the sum of the basic and acidic amino acids (22.3 per cent) is quite close to their total in the code (19.7 per cent). It is clear that natural selection counteracts the genetic code to maintain a neutral average charge in proteins by selecting Lys, Asp, and Glu, and by not

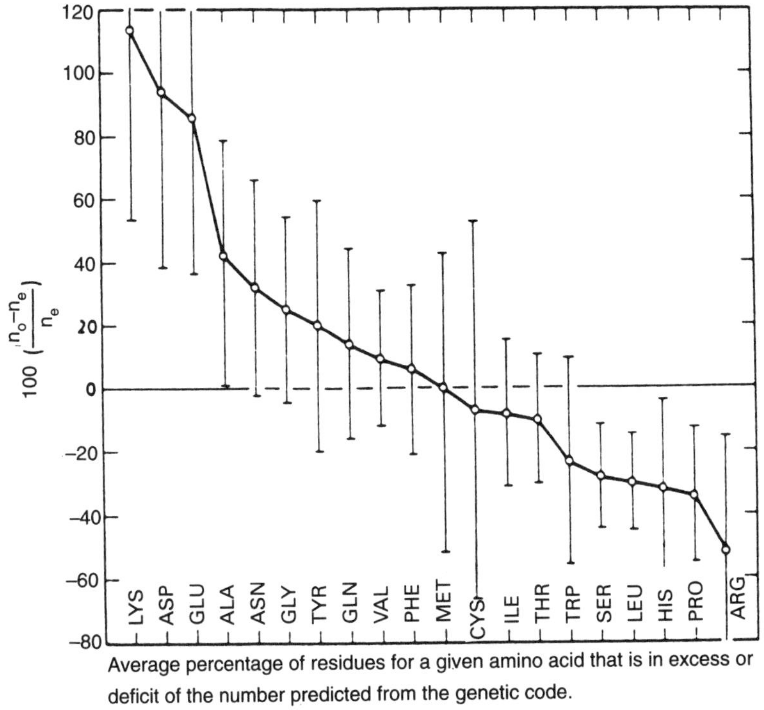

Fig. 10.2. Selection against the genetic code. The ordinate expresses the excess of the observed number n_0 of residues of each of the 20 amino acids relative to the expected number n_e in a collection of 189 proteins from 81 families. The vertical bars represent 1 standard deviation (from Holmquist, 1978).

selecting Arg (Jukes, 1973*b*; Jukes *et al.*, 1975). Jukes *et al.* (1975) suggest that the pH of the environment inhabited by primitive forms of life was higher than the present-day pH. Proteins became relatively more acidic during evolution, and the environmental pH fell, so that Asp and Glu levels are now above their genetic code levels. The Arg level fell to below its code level partly in order to decrease the basisity of proteins and partly to compensate for an increase in Lys content. Indeed, the 'underuse' of Arg codons was to some extent compensated for by 'overuse' of Lys codons (Jukes *et al.*, 1975). An alternative explanation of the 'underuse' of Arg is that Arg is an 'intruder' that displaced ornithine from an earlier code. As ornithine is similar to Lys in its structure, the 'loss' of ornithine could be compensated for by the increase of Lys (Jukes, 1973*b*). Entrance of Arg at a restricted level might be advantageous for protein functions. The Ala content in proteins is significantly higher than is expected from the genetic code, while Pro, His, Ser, and Leu are somewhat lower. Jukes (1977*b*) suggests that Ala plays a role as a 'filler' in proteins because of its small side-chain. Deviations in other amino acid contents would have been brought about by the amino acid changes

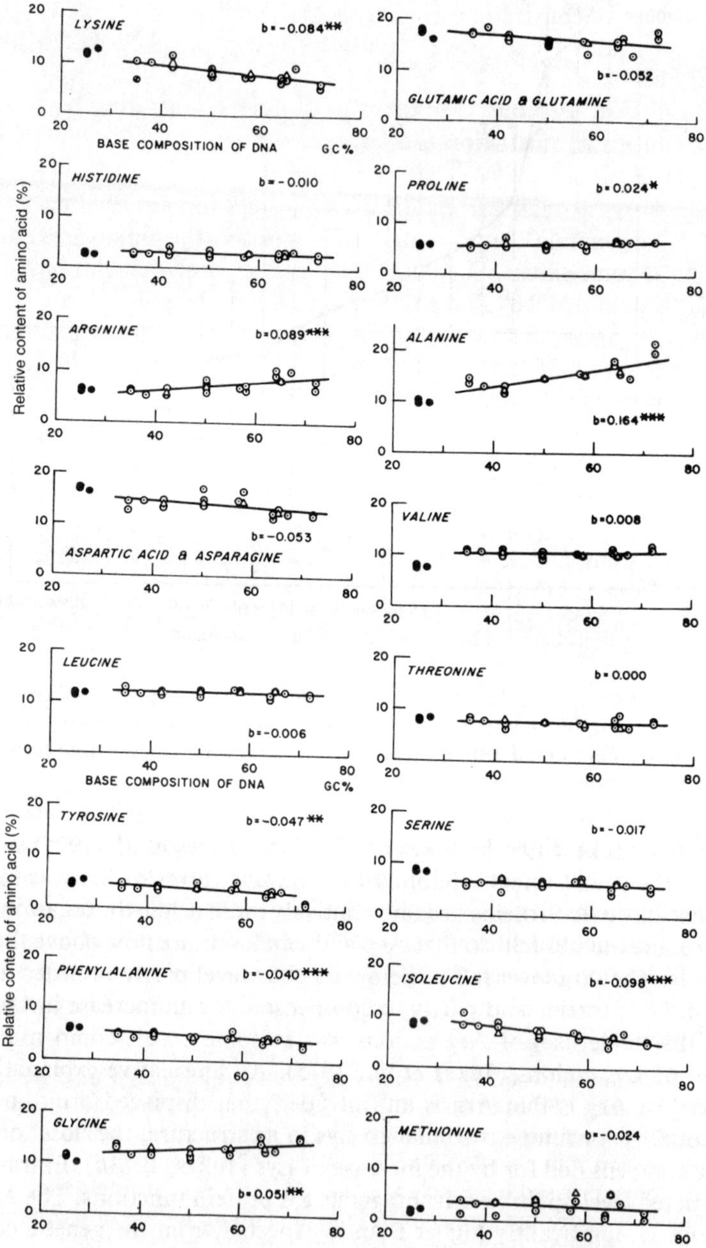

LYSINE
b = -0.084**
GLUTAMIC ACID & GLUTAMINE
b = -0.052
BASE COMPOSITION OF DNA
GC%
HISTIDINE
b = -0.010
PROLINE
b = 0.024*
Relative content of amino acid (%)
ARGININE
b = 0.089***
ALANINE
b = 0.164***
ASPARTIC ACID & ASPARAGINE
b = -0.053
VALINE
b = 0.008
LEUCINE
b = -0.006
THREONINE
b = 0.000
BASE COMPOSITION OF DNA
GC%
TYROSINE
b = -0.047**
SERINE
b = -0.017
Relative content of amino acid (%)
PHENYLALANINE
b = -0.040***
ISOLEUCINE
b = -0.098***
GLYCINE
b = 0.051**
METHIONINE
b = -0.024

sometimes affected by directional mutation pressure, and by functional selections during evolution. Ohta and Kimura (1971) noted that 'small but significant deviations from such expectation might be accounted for satisfactorily by assuming selective constraint of amino acid susbtitutions'.

10.3 Selection against the genetic code—constraint by directional mutation pressure

Figure 4.3 shows a weak correlation of the G+C content of the second codon position to the genomic G+C content, thus affecting the amino acid composition of proteins to some extent. Sueọka (1961) first explored the relationship between the G+C content of DNA and the amino acid composition of total protein in a wide variety of bacterial species and *Tetrahymena*. Ala, Gly, Arg, and Pro levels were slightly, but significantly, higher, and Tyr, Lys, and Phe levels were lower in the organisms with a high G+C content than in those with a low G+C content (Fig. 10.3).

Jukes and Bhushan (1986) noted that mitochondrial proteins were higher in Phe, Asp, and Tyr in response to AT-pressure, and that bacterial Trp synthetase was higher in Ala, Arg (CG codons), and Gly in response to GC-pressure.

Table 10.1 compares the average compositions of amino acids assigned by codons with A or U (T) at both the first and second codon positions (AT codons) and those assigned by codons with G or C at both positions (GC codons), in eight homologous ribosomal protein genes in *Micrococcus luteus* (genomic G+C=75 per cent), *E. coli* (G+C=50 per cent), and *Mycoplasma capricolum* (G+C=25 per cent) (Osawa *et al.*, 1992). The total content of Phe, Ile, Met, Tyr, Asn, and Lys assigned by AT codons increases significantly with increasing genomic A+T content, and amino acids assigned by GC codons, Pro, Ala, Arg (CGN), and Gly tend to increase with increasing genomic G+C. The *b/a* values are the most marked findings in Table 10.1. The results are essentially the same as Sueoka's (1961), proving that directional mutation pressure affects the amino acid composition of proteins. Its effect is presumably exerted mostly on amino acids in neutral or near-neutral sites in the protein molecules. A slight but significantly positive correlation between the G+C content of the codon second position and the non-silent fraction of the first position with the genomic G+C content is the

---

Fig. 10.3. Relation between DNA base composition and amino acid composition of protein in bacteria. The relative molar content of each amino acid is expressed as a percentage of the amino acid content over the sum of stable amino acids: Lys, His, Arg, Asp plus Asn, Glu plus Gln, Pro, Ala, Val, Leu, Tyr, and Phe. The base composition of DNA is expressed as a molar content (%) of G + C. b, regression co-efficient. *Tetrahymena* data are not included in the calculation. Levels of significance of the regression co-efficient (i.e. the slope) are shown by asterisks. * Significant ($0.02<P<0.05$); ** highly significant ($0.001<P<0.02$); *** extremely significant ($P<0.001$); ● *Tetrahymena pyriformis*; ⊙ cultured in enriched broth medium; △ cultured in minimal medium (from Sueoka, 1961).

Table 10.1 Composition of amino acids assigned by AT and GC codons

Amino acid[a]	Molar content (%) in[b] *M. luteus* (G+C, 75%)	*E. coli* (G+C, 50%)	*M. capricolum* (G+C, 25%)
AT codons			
Phe	3.0	3.8	4.2
Ile	5.8	7.4	10.3
Met	2.3	2.8	2.0
Tyr	2.2	2.2	3.2
Asn	2.6	3.1	5.3
Lys	6.1	8.2	11.8
Total (a)	22.0	27.5	35.8
GC codons			
Pro	4.2	3.7	3.4
Ala	9.0	10.0	6.2
Arg (CGN)	7.4	7.2	1.0
Gly	10.0	9.7	7.6
Total (b)	30.6	30.6	18.2
(b)/(a)	1.39	1.15	0.51

[a] AT or GC refers to codons with AA, AT, TA, and TT, or GG, GC, CG, and CC in the first two nucleotide positions.

[b] The values are molar content (%) of each amino acid in the protein genes in the *spc* operon of *Micrococcus luteus* (Ohama *et al.*, 1990*a*), *E. coli* (Cerretti *et al.*, 1983), and *Mycoplasma capricolum* (Ohkubo *et al.*, 1987) (from Osawa *et al.*, 1992).

result of this directional amino acid substitution. Note that the extent of amino acid substitutions affected by directional mutation pressure is considerably smaller than the extent of those substitutions affected by functional selection. This situation may be seen in the following examples: The amino acid composition coded by the *str* operon in *M. luteus* is almost identical to the corresponding composition in *E. coli* (Table 10.2). In this case, the difference in G+C content between these two species is almost entirely attributable to silent substitutions, except that *M. luteus* has apparently replaced a small number of AUU Ile codons in *E. coli* with Val. The protein composition of the *str* (*rps* 5) gene in *E. coli* is more similar to that in *M. luteus* than it is to the average composition of 199 *E. coli* genes, despite the great difference in the G+C content in the silent sites between *E. coli* and *M. luteus*. Thus, the *str* operon did not respond appreciably to GC-pressure in the replacement of amino acids, probably because it is highly conserved.

The combined effects of GC-pressure and compositional non-randomness are shown in Table 10.3. In first codon positions, all three sequences are high in G as a result of high contents of Asp, Glu, Ala, and Gly, also Val in *M. luteus* and

Table 10.2 Codon usage in the *str* operon of *Micrococcus luteus* and *Escherichia coli*

	M. luteus	*E. coli*		*M. luteus*	*E. coli*		*M. luteus*	*E. coli*		*M. luteus*	*E. coli*
Phe (UUU)	0	7	Ser (UCU)	0	27	Tyr (UAU)	0	8	Cys (UGU)	0	4
Phe (UUC)	43	37	Ser (UCC)	36	18	Tyr (UAC)	37	28	Cys (UGC)	6	6
Leu (UUA)	0	2	Ser (UCA)	1	3	Stop (UAA)	1	3	Stop (UGA)	3	1
Leu (UUG)	0	2	Ser (UCG)	16	0	Stop (UAG)	0	0	Trp (UGG)	8	8
Leu (CUU)	1	4	Pro (CCU)	5	6	His (CAU)	1	7	Arg (CGU)	23	67
Leu (CUC)	35	5	Pro (CCC)	27	2	His (CAC)	30	24	Arg (CGC)	61	16
Leu (CUA)	0	0	Pro (CCA)	0	5	Gln (CAA)	0	1	Arg (CGA)	1	0
Leu (CUG)	54	78	Pro (CCC)	35	52	Gln (CAG)	47	39	Arg (CGG)	2	1
Ile (AUU)	1	14	Thr (ACU)	3	34	Asn (AAU)	3	2	Ser (AGU)	0	6
Ile (AUC)	73	72	Thr (ACC)	68	40	Asn (AAC)	43	41	Ser (AGC)	1	3
Ile (AUA)	0	0	Thr (ACA)	0	5	Lys (AAA)	0	69	Arg (AGA)	0	0
Met (AUG)	38(1)[a]	36(3)[a]	Thr (ACG)	24	0	Lys (AAG)	89	20	Arg (AGG)	3	0
Val (GUU)	1	70	Ala (GCU)	6	48	Asp (GAU)	5	16	Gly (GGU)	15	76
Val (GUC)	67	1	Ala (GCC)	83	6	Asp (GAC)	71	55	Gly (GGC)	100	38
Val (GUA)	0	39	Ala (GCA)	3	25	Glu (GAA)	0	90	Gly (GGA)	0	2
Val (GUG)	74(3)[a]	12(1)[a]	Ala (GCG)	21	35	Glu (GAG)	113	21	Gly (GGG)	4	2

[a] Initiation codons.
From Osawa *et al.* (1987).

Table 10.3 Nucleotide percentages in three codon positions

Codon position and protein	Nucleotides (%) U	C	A	G
First position				
consensus protein	18.5	18.2	28.5	34.8
M. luteus, str	10.7	23.4	25.1	40.9
E. coli, str	11.2	23.0	25.6	40.2
Second position				
consensus protein	25.3	23.7	33.0	18.0
M. luteus, str	28.1	23.8	31.9	16.3
E. coli, str	28.4	22.9	31.5	17.2
Third position				
consensus protein	25.9	25.9	23.3	24.8
M. luteus, str	4.6	56.7	0.4	38.3
E. coli, str	29.7	29.4	18.1	22.9

Calculated for a protein of average composition (consensus protein) as compiled by Jukes *et al.* (1975) and for *str* operons in *M. luteus* and *E. coli* (see Table 10.2).
From Osawa *et al.* (1987).

E. coli. *M. luteus* and *E. coli* are low in T because of low levels of Cys, and Ser. Second codon positions are high in A because of high levels of Lys, Asp, and Glu, and low in G because of low levels of Cys, Trp, and Arg. *M. luteus* and *E. coli* show close agreement in the nucleotide composition of all first and second positions because of the great similarity in amino acid composition of *str* (ribosomal protein S5) proteins in these two species. The higher levels of Ala, Lys, Asp, Glu, and Gly, and the lower levels of Cys, Trp, and Arg than would be expected from their number of codons in the genetic code agree well with the levels shown in Fig. 10.2. Third codon positions are very high in G+C in *M. luteus* because of GC-pressure and predominance of CNN anticodons.

10.4 Amino acid usage in proteins affects the levels of tRNAs

We have seen that the amino acid composition of protein is determined primarily by the composition of the genetic code and is modified by various evolutionary constraints. It is therefore expected that the number of intracellular tRNAs is related to the number of codons to be translated (roughly speaking, the number of amino acids in proteins). In fact, there is a linear correlation between the relative amounts of all isoacceptors for each amino acid and the amino acid composition of proteins. As shown in Fig. 10.4(a), the higher the content of an

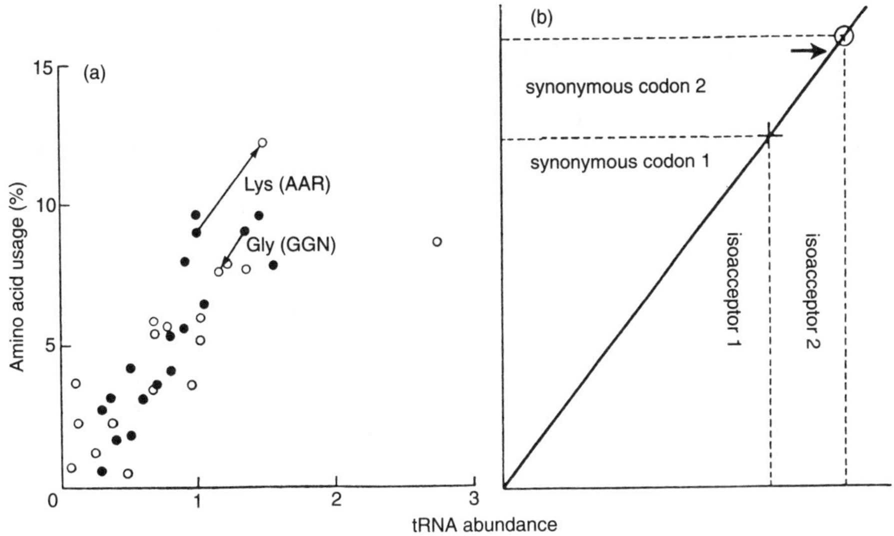

Fig. 10.4. Co-linearity between amino acid usage and tRNA content. (a) Amino acid compositions of ribosomal proteins of *Mycoplasma capricolum* (○) and *E. coli* (●) are plotted against the relative levels of tRNAs for the amino acid. The amino acid composition of ribosomal proteins is almost the same as that of total proteins (*E. coli*). Arrow for Lys shows a shift from *E. coli* to *M. capricolum*; arrow for Gly shows a shift from *M. capricolum* to *E. coli* (see text). The amount of tRNAs^{Ile} of *M. capricolum* is exceptionally high; the reason for this is unknown. (b) Schematic representation of the relationship between synonymous codons and their corresponding tRNAs (for details, see text) (redrawn from Yamao *et al.* (1991), courtesy of F. Yamao).

amino acid in the proteins, the higher the content of the tRNA(s) that translates the codons for the amino acid in both *Mycoplasma capricolum* and *E. coli* (Yamao *et al.*, 1991). Thus, the amount of tRNAs for each amino acid seems to be regulated to a particular level in parallel with the requirement of the amino acid for total protein synthesis.

As most proteins are conservative in their amino acid sequences, the patterns of amino acid usage, at least among eubacteria, do not differ much. The fact that the levels of tRNAs for each amino acid are correlated with amino acid usage implies that the pattern of tRNA levels for the amino acid is invariable among eubacteria. In fact, both amino acid composition and the pattern of the relative amount of tRNA for amino acid in *E. coli* and *M. capricolum* are similar. This is in strong contrast to the codon usage patterns (choice among synonymous codons) of eubacteria, and hence the pattern of relative amounts of isoacceptor tRNAs differ from one species to another, as discussed in Section 5.1.

Figure 10.4(b) shows the quantitative relationships between total tRNA, amino acid, synonymous codon, and isoacceptor tRNA. For example, taking Lys, point (○) may be taken to indicate the Lys content of proteins and the total amount of $tRNAs^{Lys}$. The position of this point is species-independent, at least in eubacteria. Synonymous codon 1 (AAA) is mostly translated by isoacceptor 1 ($tRNA^{Lys}_{UUU}$) and codon 2 (AAG) by isoacceptor 2 ($tRNA^{Lys}_{CUU}$) in *E. coli* (point +), while in *M. capricolum* this point (+) shifts up, as indicated by the arrow, because there is almost no usage of UUG and hence a low level of $tRNA^{Lys}_{CUU}$. Thus, the amount of isoacceptor tRNA is determined by the amount of the corresponding codon in a species-specific manner.

Although the amino acid composition in *M. capricolum* and *E. coli* proteins are very similar, there are minor but significant differences in composition of some amino acids, as noted in Section 10.3. Ala, Arg, Gly, and Val showed higher scores in *E. coli* than in *M. capricolum* and, conversely Asn, Ile, Lys, and Ser are lower, because of differences in the G + C content of their genomes. Even though the G + C content-dependent difference in the amino acid composition is minor, tRNA levels seem to have adapted to it. Most notably, for example, Lys (AAR; mainly AAA) has a greater usage in *M. capricolum* than in *E. coli*, and Gly (GGN) and Ala (GCN) a lesser one. The relative level of $tRNAs^{Lys}$ in *M. capricolum* is significantly higher than that in *E. coli*, and those of Gly and Ala tRNAs are lower (Fig. 10.4(a)).

Epilogue

In some recent textbooks, non-universal genetic codes, such as UGA Trp in *Mycoplasma capricolum* and UAR Gln in ciliated protozoans, are dealt with simply as exceptions. Such treatment gives the impression that nearly all organisms use the universal genetic code. But is this true? Despite an enormous diversity of organisms, all of which are derived from a single ancestor, until recently only a handful of standard organisms, such as *Escherichia coli*, *Bacillus subtilis*, *Saccharomyces cerevisiae*, *Drosophila* spp., and vertebrates, had been examined genetically. With the development of molecular phylogenetic studies and the rapid progress of gene technology, interest has begun to focus on various odd organisms. As a result, a relatively high incidence of non-universal codes has been discovered. Codon UGA Trp has been found in eight *Mycoplasma* species and related bacteria; at least two kinds of non-universal genetic code are used independently in ciliated protozoans; the same code change was found in two different organismic lines—ciliated prozoans and unicellular green algae—a yeast line uses yet another code. All the non-plant mitochondria that have been examined, except oömycetes and slime moulds use non-universal genetic codes, which are more or less characteristic for each line. It is remarkable that mitochondria from one species use more than two non-universal codons; six in yeasts, four or five in many invertebrates, and four in vertebrates. Thus, non-universal genetic codes are widely distributed in various groups of organisms and organelles (Fig. 11.1).

Some of the principles governing the evolution of the genetic code, upon which the code capture theory is based, have come to light in organisms with very high genomic A + T or G + C contents. The existence of unassigned codons has been demonstrated in these organisms, and mechanisms of the code change for UGA stop to Trp, and of some other changes, have been uncovered. It should be stressed that such laboratory data would never have been obtained with the standard organisms alone.

Independent occurrence of the same code change in more than one different organismic or mitochondrial line suggests that the changes take place by essentially the same mechanisms throughout all nuclear and mitochondrial systems.

The non-universal genetic codes are not produced randomly, but are derived from the universal genetic code as the result of a series of non-disruptive changes. As discussed in detail above, many codons have their assignments changed by disappearing from mRNA sequences and then reappearing by codon capture. The codon capture is driven by directional mutation pressure, modification of RF, or genomic economization. This produces an unassigned codon, which can be reassigned upon appearance of a new tRNA or RF that recognizes the codon.

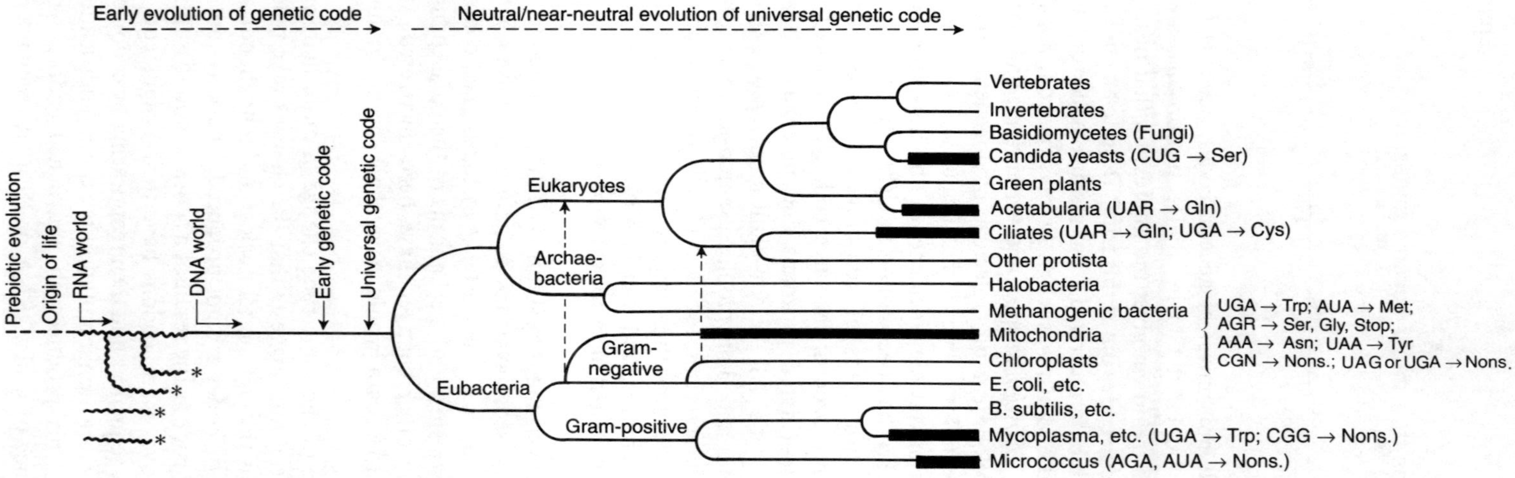

Fig. 11.1. Phylogenetic representation of evolution of the genetic code. The branching lengths are diagrammatic and not to scale. -----> endosymbiosis; ★ extinction.

Table 11.1 Codons that might become unassigned

UUU Phe	UCU Ser	UAU Tyr	UGU Cys
UUC Phe	UCC Ser	UAC Tyr	UGC Cys
UUA Leu	**UCA Ser**	**UAA Stop***	**UGA Stop***
UUG Leu	UCG Ser	**UAG Stop***	UGG Trp
CUU Leu*	CCU Pro	CAU His	**CGU Arg***
CUC Leu*	CCC Pro	CAC His	**CGC Arg***
CUA Leu*	**CCA Pro**	**CAA Gln**	**CGA Arg***
CUG Leu*	CCG Pro	CAG Gln	**CGG Arg***
AUU Ile	ACU Thr	AAU Asn	AGU Ser
AUC Ile	ACC Thr	AAC Asn	AGC Ser
AUA Ile*	**ACA Thr**	**AAA Lys***	**AGA Arg***
AUG Met	ACG Thr	AAG Lys	**AGG Arg***
GUU Val	GCU Ala	GAU Asp	GGU Gly
GUC Val	GCC Ala	GAC Asp	GGC Gly
GUA Val	**GCA Ala**	**GAA Glu**	**GGA Gly**
GUG Val	GCG Ala	GAG Glu	GGG Gly

Candidates for unassigned codons are shown in bold.
* Unassigned and/or changed codons that have been reported.

This means that codons that may be unassigned would have the potential for reassignment (Table 11.1).

New changes will be discovered as more organisms or organelles are examined. A step towards finding new changes in nuclear genetic codes would be to discover organisms with high genomic G+C or A+C content, or a group of organisms with a wide range of genomic G+C contents, such as *Candida* spp. Indeed, all the organisms in which changes are found carry such genomes. As mitochondrial genomes are under more intense evolutionary pressures (such as higher mutation rates, genomic economization, and directional mutation pressure) than nuclear genomes, the incidence for code changes will be higher. All these new changes, both in nuclear and mitochondrial systems, could be found from the candidates shown in Table 11.1, in accordance with the codon capture theory. However, it is difficult to show the complete list of the expected amino acid species for the codons upon reassignment—there are too many possibilities. As discussed above, the reassignment of a codon may be implemented by a change in the anticodon, a change of tRNA identity, or by a change in codon–anticodon pairing. Indeed, there are various ways to change the anticodon—single mutation, base modification, or creation of a new anticodon by splicing, as is the case with *Candida* $tRNA^{Ser}$ (see Section 6.5.4). This last event had not even been speculated upon before it was discovered. Nevertheless, within a certain sphere it is possible to deduce which amino acid may be specified upon reassignment. Such predictions have been made since 1988, when the codon capture theory was proposed (Osawa and Jukes, 1989). These changes involve:

UUR: Leu to Phe, Ser or Val pairing with anticodon UAA formed by mutation of GAA (Phe), UGA (Ser) or UAC (Val).

AGA and AGG: Arg to Ser or Gly pairing with anticodon UCU by mutation from GCU (Ser) or UCC (Gly).

CGG: Arg to Trp pairing with anticodon CCG by mutation from CCA (Trp).

UAA and UAG: stop to Glu, Ser or Tyr pairing with anticodon UUA by mutation from UUC (Glu), UGA (Ser) or GUA (Tyr).

For all these changes, it was postulated that a codon originally for amino acid A but to be reassigned to B became unassigned. The codon would then be created from other synonymous codons for B by a single mutation and finally would be captured by a new tRNA for B in which a single mutation had occurred on the anticodon in a duplicate of the pre-existing tRNA molecule. This view is supported strongly by recent evidence that specific aminoacylation of tRNAs can be independent from anticodons in many cases (Schimmel *et al.*, 1993). The base modification and the tRNA identity change upon mutation of the tRNA were not considered. The change of AGA and AGG from Arg to Ser has been known since 1987, and the change of AGA and AGG from Arg to Gly was first discovered in ascidian mitochondria in 1993. Codon UAA from stop to Tyr was reported in planarian mitochondria in 1992, although the mechanisms of change would involve tRNA modification rather than mutation of the anticodon (see below). Note that the reported code changes are not restricted to those between codons specifying chemically related amino acids, as exemplified by changes of AGR from Arg to Gly, AGR from Arg to Ser, or AAA from Lys to Asn.

The discovery of the last change, Lys (AAA) to Asn, prompted the formulation of one type of the code changes for NNA codons in 2-codon sets. The change of AAA Lys to Asn was supposed to include steps in which AAA first became unassigned by converting to AAG Lys pairing with anticodon CUU, and by the disappearance of the *UUU anticodon (remember that all NAA codons may be unassigned). The first anticodon nucleoside of $tRNA^{Asn}_{GUU}$ pairing with AAY codons then became modified to I or equivalent (*G) so as to expand the wobbling capacity and enable recognition of AAA codon as Asn (see Table 6.6). The duplication of the gene for $tRNA^{Asn}_{GUU}$ is not necessary. By the same token, codon UUA (Leu) might change to specify Phe pairing with the newly emerged $tRNA^{Phe}_{*GAA}$. Similarly, UAA might change from stop to Tyr, UGA from stop to Cys, CAA from Gln to His, GAA from Glu to Asp, AGA from Arg to Ser, and, in mitochondria, AUA from Met to Ile as a reversal of Ile to Met. Some of these changes, in addition to AAA Lys to Asn, indeed took place: UGA stop to Cys in *Euplotes*, AGA Arg to Ser in most invertebrate mitochondria, AUA Met to Ile in echinoderm and planarian mitochondria and UAA stop to Tyr in planarian mitochondria.

All the code changes can proceed in the reverse direction. Indeed, the changes of AUA Ile to Met in mitochondria and of CUG Leu to Ser in *Candida* spp. reversed as described above. Thus, the genetic code seems to possess considerable flexibility with respect to changes in the amino acid (or stop) assignment of codons.

Schultz and Yarus (1994) propose that reassignment of codons in the genetic code is 'facilitated by a translationally ambiguous intermediate'. Their proposal is that 'codons have been reassigned via intermediates in which the transitional codon has two meanings. The simultaneous reading of the transitional codon by two tRNAs, one cognate and the other non-cognate . . . is the crucial intermediate step', in contrast to the production of unassigned codon in the theory of Osawa and Jukes (1989). Schultz and Yarus say 'in successful reassignments, the near-cognate tRNA gains function at the codon in transition while cognate tRNA loses function'. They emphasize that mutations at base pair 27–43 at the top of the anticodon 'facilitate normally forbidden near-cognate ribosomal pairings between codon and anticodon', so that reading of a cognate codon 'by a non-cognate tRNA therefore is possible'.

Let us take, for example, the reassignment of AAA from Lys to Asn and adapt it to the Schultz and Yarus model. The function of cognate $tRNA^{Lys}_{UUU}$ for codon AAA can be diminished by mutation, although Schultz and Yarus do not mention the driving force of such mutations. If near-cognate $tRNA^{Asn}_{GUU}$ emerges, AAA may be read as Asn (by near-cognate $tRNA^{Asn}_{GUU}$) or as Lys (by $tRNA^{Lys}_{UUU}$). For this to occur, the $tRNA^{Asn}_{GUU}$ becomes a new near-cognate $tRNA^{Asn}_{GUU}$ by mutation of nucleotides outside the anticodon (such as those at base pair 27–43), so that the tRNA can read both AAY and AAA as Asn. Positive selection would then operate to facilitate the reading of AAA as Asn by the new $tRNA^{Asn}_{GUU}$, while dysfunctional $tRNA^{Lys}$ loses function. Note, however, that all AAAs in the original gene (or mRNA) sequences were at Lys sites, and that these were not the 'transitional' codons. Reassignment of AAA codons to Asn would occur only at the Asn sites originally occupied by AAY Asn codons that subsequently mutated to AAA. In general, at the initial stage of reassignment the codons to be reassigned (in this case AAA, to be formed from AAY) would be much fewer in number than the synonymous codons (AAY) that have not undergone reassignment, and than the codons that should not have been subjected to reassignment (AAA at Lys site). Ambiguous readings of all AAA codons at Lys sites are obviously deleterious for an organism. The whole process could proceed only when there are very small numbers of AAAs at Lys sites in the functional sites of the original gene sequences. If this is not the case, and sometimes even when it is, the presence of ambiguous codons would be more disadvantageous (or at best slightly deleterious) than their absence in the natural population, implying that such an organism would have more chance of negative selection. In any case, AAAs at Lys sites must finally disappear by converting to AAG Lys, which is read by $tRNA^{Lys}_{CUU}$, as stressed by Osawa and Jukes (1989). Directional mutation pressure would play a decisive role in reducing the number of AAA Lys codons, especially in prokaryotes and some mitochondria. The importance of specificity of coding for survival must be emphasized.

Schultz and Yarus state that ambiguous codon assignment is well established. For example, UGA encodes stop, Trp, and Sec in wild-type *E. coli*. Sec is incorporated at UGA codons in a few *E. coli* proteins by a special and elaborate mechanism that keeps internal UGA codons separated from terminal UGA codons (Böck *et al.*, 1991). The 1562 genes in wild-type *E. coli* have a total of 6712 UGG Trp codons and 417 UGA stop codons, so there is no indication that UGA usually encodes Trp in wild-type *E. coli*. They also argue that 'stop codons are easily read by nonsense suppressor tRNAs in non-transitional cells. Reading by a near-cognate tRNA therefore is probable'. Here again, one must remember that the target codons to be reassigned are not at the termination sites but in the reading frames (in this case, mostly UGG). In contrast, the Osawa–Jukes model (see Fig. 6.18) includes the functions of release factors (RF) (Jukes *et al.*, 1991). These are not mentioned by Schultz and Yarus. No ambiguity occurs in the Osawa–Jukes proposal for change of codon assignment.

Schultz and Yarus stressed that new codon meanings do not occur at random and take place in an ordered fashion, as those for Asn, Tyr, Ser, and Cys require codon–anticodon pairings between G and A in AAA–GUU, UAA–GUA, AGA–GCU, and UGA–GCA codon–anticodon pairs, respectively. These require G-A mispairings at the third codon position, according to Schultz and Yarus, caused from mutations of, for example, base pair 27–43. But, as discussed above, Osawa *et al.*'s (1992) proposal was that G was converted to I or its equivalent (*G; modified G), which pairs with A, in addition to U and C, by standard wobble rules (Crick, 1966*a*).

Schultz and Yarus say that the 'Osawa–Jukes theory apparently predicts that all tRNAs would have an opportunity to capture the unassigned codon after its removal from the genome'. This is incorrect; only tRNAs with closely similar anticodons have such opportunities. Also, possible codon candidates for reassignment are limited by other factors (see Section 5 and Table 11.1).

All the codon reassignments that have been reported can be explained by Osawa and Jukes' codon capture theory without considering ambiguous intermediates. Widespread occurrence of unassigned codons throughout mitochondria and organisms (see Table 5.3) is consistent with the codon capture theory.

The use of UGA as a Sec codon is one of the active fields in the study of evolution of the code, and is still under much discussion. Generally, a codon cannot have two simultaneous meanings. UGA is a remarkable exception; in-frame UGA is read as Sec, and UGA at the termination site functions as a stop codon in an organism. In some organisms, UGA is used as a codon for Trp or Cys. Is Sec a relic of the 'anaerobic world', or is it a recent evolutionary improvement?

RNA editing is another important field, and its evolutionary meaning is becoming clearer, as described above. However, the molecular mechanisms of substitution RNA editing are still quite obscure and will hopefully be explored in the near future.

Current knowledge, as described in this book, has made possible the deduction of the evolution of the genetic code from the progenote cells to various extant organisms and organelles. However, the origin and early evolution of the genetic code, and so the origin of life itself, still pose an enormous challenge.

References

Ainley, W. M., Macreadie, I. G., and Butowa, R. A. (1985). *var1* gene on the mitochondrial genome of *Torulopsis glabrata*. *J. Mol. Biol.*, **184**, 565–76.

Akiyama, M., Maki, H., Sekiguchi, M., and Horiuchi, T. (1989). A specific role of *Mut*T protein to prevent dG·dA mispairing in the DNA replication. *Proc. Natl. Acad. Sci. USA*, **86**, 3949–52.

Andachi, Y., Yamao, F., Iwami, M., Muto, A., and Osawa, S. (1987). Occurrence of unmodified adenine and uracil at the first position of anticodon in threonine tRNAs in *Mycoplasma capricolum*. *Proc. Natl. Acad. Sci. USA*, **84**, 7398–402.

Andachi, Y., Yamao, F., Muto, A., and Osawa, S. (1989). Codon recognition patterns as deduced from sequences of the complete set of transfer RNA species in *Mycoplasma capricolum*: resemblance to mitochondria. *J. Mol. Biol.*, **209**, 37–54.

Anderson, S., Bankier, A. T., Barrell, B. G., de Bruijn, M. H. L., Coulson, A. R., Drouin, J., *et al.* (1981). Sequence and organization of the human mitochondrial genome. *Nature*, **290**, 457–65.

Anderson, S., de Bruijn, M. H. L., Coulson, A. R., Eperon, I. C., Sanger, F., and Young, I. G. (1982). Complete sequence of bovine mitochondrial DNA: conserved feature of the mammalian mitochondrial genome. *J. Mol. Biol.*, **156**, 683–717.

Angata, K., Kuroe, K., Yanagisawa, K., and Tanaka, Y. (1995). Codon usage, genetic code and phylogeny of *Dictyostelium discoideum* mitochondrial DNA as deduced from a 7.3-kb region. *Curr. Genet.*, **27**, 249–56.

Aota, S.-I. and Ikemura, T. (1986). Diversity in G+C content at the third position of codons in vertebrate genes and its cause. *Nucleic Acids Res.*, **14**, 6345–55.

Araki, T., Asakawa, S., Kumazawa, Y., Miura, K.-I., and Watanabe, K. (1988). Gene organization of tRNAs and their secondary structures found in starfish mitochondrial genome. *Nucleic Acids Res., Symp. Ser.*, **20**, 93–4.

Asahara, H., Himeno, H., Tamura, K., Nameki, N., Hasegawa, T., and Shimizu, M. (1994). *Escherichia coli* seryl-tRNA synthetase recognizes $tRNA^{Ser}$ by its characteristic tertiary structure. *J. Mol. Biol.*, **236**, 738–48.

Asakawa, S., Kumazawa, Y., Araki, H., Himeno, T., Miura, K., and Watanabe, K. (1991). Strand-specific nucleotide composition bias in echinoderm and vertebrate mitochondrial genomes. *J. Mol. Evol.*, **32**, 511–20.

Baer, R. J. and Dubin, D. T. (1980). The sequence of a possible 5S RNA-equivalent in hamster mitochondria. *Nucleic Acids Res.*, **8**, 3603–10.

Barahona, I., Soares, H., Cyrne, L., Penque, D., Denoulet, P., and Rodrigues-Pousada, C. (1988). Sequence of one alpha- and two beta-tubulin genes of *Tetrahymena pyriformis*. *J. Mol. Biol.*, **202**, 365–82.

Baron, C. and Böck, A. (1991). The length of the aminoacyl-acceptor stem of the selenocysteine-specific $tRNA^{Sec}$ of *Escherichia coli* is the determinant for binding to elongation factors SELB or Tu. *J. Biol. Chem.*, **266**, 20375–9.

Baron, C., Heider, J., and Böck, A. (1990). Mutagenesis of *sel*C, the gene for the selenocysteine-inserting tRNA-species in *E. coli*: effects on *in vivo* function. *Nucleic Acids Res.*, **18**, 6761–6.

Baron, C., Westhof, E., Böck, A., and Giegé, R. (1993). Solution structure of selenocysteine-inserting tRNA^{Sec} from *Escherichia coli*. Comparison with canonical tRNA^{Ser}. *J. Mol. Biol.*, **231**, 274–92.

Barrell, B. G., Bankier, A. T., and Drouin, J. (1979). A different genetic code in human mitochondria. *Nature*, **282**, 189–94.

Barrell, B. G., Anderson, S., Bankier, A. T., de Bruijn, M. H. L., Chen, E., Coulson, A. R., *et al.* (1980). Different pattern of codon recognition by mammalian mitochondrial tRNAs. *Proc. Natl. Acad. Sci. USA*, **77**, 3164–6.

Bass, B. L. (1993). RNA editing: new uses for old players in the RNA world. In *The RNA world*, (ed. Gesteland, R. F. and Atkins, J. F.), pp. 383–418, Cold Spring Harbor Laboratory Press, Cold Spring Harbor, New York.

Begu, D., Graves, P. V., Domec, C., Arselin, G., Litvak, S., and Araya, A. (1990). RNA editing of 2 wheat mitochondrial ATP synthase subunit 9: direct protein and cDNA sequencing. *The Plant Cell*, **2**, 1283–90.

Bernardi, G. (1989). The isochore organization of the human genome. *Ann. Rev. Genet.*, **23**, 637–61.

Bernardi, G. and Bernardi, G. (1986). Compositional constraints and genome evolution. *J. Mol. Evol.*, **24**, 1–11.

Bernardi, G., Olofsson, B., Filipski, J., Zerial, M., Salinas, J., Cuny, G., *et al.* (1985). The mosaic genome of warm-blooded vertebrates. *Science*, **228**, 953–8.

Berry, M. J., Banu, L., and Larsen, P. R. (1991*a*). Type 1 iodothyronine deiodinase is a selenocysteine-containing enzyme. *Nature*, **349**, 438–40.

Berry, M. J., Banu, L., Chen, Y., Mandel, S. J., Kieffer, J. D., Harney, J. W., and Larsen, P. R. (1991*b*). Recognition of UGA as a selenocysteine codon in Type 1 deiodinase requires sequences in the 3′ untranslated region. *Nature*, **353**, 273–6.

Berry, M. J., Banu, L., Harney, J. W., and Larsen, P. R. (1993). Functional characterization of the eukaryotic SECIS elements which direct selenocysteine insertion at UGA codons. *EMBO J.*, **12**, 3315–22.

Bessho, Y., Ohama, T., and Osawa, S. (1991). Planarian mitochondria II: the unique genetic code as deduced from COI gene sequences. *J. Mol. Evol.*, **34**, 324–30.

Bibb, M. J., VanEtten, R. A., Wright, C. T., Walberg, M. W., and Clayton, D. A. (1981). Sequence and gene organization of mouse mitochondrial DNA. *Cell*, **26**, 167–80.

Binder, S., Marchfelder, A., Brenicke, A., and Wissinger, B. (1992). RNA editing in trans-splicing intron sequences of *nad* 2 mRNAs in *Oenothera* mitochondria. *J. Biol. Chem.*, **267**, 7615–23.

Björk, G. R. (1987). Modification of stable RNA. In *Escherichia coli and Salmonella typhimurium: Cellular and molecular biology*, (ed. Neidhardt, F. C., Ingraham, J. L., Low, K. B., Magasanik, B., Schaechter, M., and Umbarger, H. E.), Vol. 1, pp. 719–31. American Society of Microbiology, Washington, D.C.

Björk, G. R., Ericson, J. U., Gustafsson, C. E. D., Hagervall, T. G., Jönsson, Y. H., and Wikström, P. M. (1987). Transfer RNA modification. *Ann. Rev. Biochem.*, **56**, 263–87.

Böck, A. and Stadtman, T. C. (1988). Selenocysteine, a highly specific component of certain enzymes, is incorporated by a UGA-directed co-translational mechanism. *BioFactors*, **1**, 245–50.

Böck, A., Forchhammer, K., Heider, J., Leinfelder, W., Sawers, G., Veprek, B., and Zinoni, F. (1991). Selenocysteine: the 21st amino acid. *Mol. Microbiol.*, **5**, 515–20.

Bonitz, S. G. and Tzagoloff, A. (1980). Assembly of the mitochondrial membrane system. Sequences of yeast mitochondrial tRNA genes. *J. Biol. Chem.*, **255**, 9075–81.

Bonitz, S. G., Berlani, R., Coruzzi, G., Li, M., Macino, G., Nobrega, F. G., *et al.* (1980). Codon recognition rules in yeast mitochondria. *Proc. Natl. Acad. Sci. USA*, **77**, 3167–70.

Boer, P. H. and Gray, M. W. (1988). Transfer RNA genes and the genetic code in *Chlamydomonas reinhardtii* mitochondria. *Curr. Genet.*, **14**, 583–90.

Borén, T., Elias, P., Samuelsson, T., Claesson, C., Barciszewska, M., Gehrke, C. W., Kuo, K. C., and Lustig, F. (1993). Undiscriminating codon reading with adenosine in the wobble position. *J. Mol. Biol.*, **230**, 739–49.

Boyen, C., Leblanc, C., Bonnard, G., Grienenberger, J.-M., and Kloareg, B. (1994). Nucleotide sequence of the cox3 gene from *Chondrus crispus*: evidence that UGA encodes tryptophan and evolutionary implications. *Nucleic Acids Res.*, **22**, 1400–3.

Brachet, J. (1944). *Embryologie chimique*, p. 510. Masson et Cie., Paris.

Brenner, S. (1957). On the impossibility of all overlapping triplet codes in information transfer from nucleic acid to proteins. *Proc. Natl. Acad. Sci. USA*, **43**, 687–94.

Brenner, S., Jacob, F., and Meselson, M. (1961). An unstable intermediate carrying information from genes to ribosomes for protein synthesis. *Nature*, **190**, 576–81.

Brenner, S., Stretton, A. O. W., and Kaplan, S. (1965). Genetic code: the 'nonsense' triplets for chain termination and their suppression. *Nature*, **206**, 994–8.

Browing, K. S. and RajBhandary, U. L. (1982). Cytochrome oxidase subunit III gene in *Neurospora crassa* mitochondria: location and sequence. *J. Biol. Chem.*, **257**, 5253–6.

Brown, C. M., Stockwell, P. A., Trotman, C. N. A., and Tate, W. P. (1990*a*). The signal for the termination of protein synthesis in procaryotes. *Nucleic Acids Res.*, **18**, 2079–86.

Brown, C. M., Stockwell, P. A., Trotman, C. N. A., and Tate, W. P. (1990*b*). Sequence analysis suggests that tetra-nucleotides signal termination of protein synthesis in eukaryotes. *Nucleic Acids Res.*, **18**, 6339–45.

Burk, R. F. and Hill, K. E. (1993). Regulation of selenoproteins. *Ann. Rev. Nutr.*, **13**, 65–81.

Canaday, J., Dirheimer, G., and Martin, R. P. (1980). Yeast mitochondrial methionine initiator tRNA: characterization and nucleotide sequence. *Nucleic Acids Res.*, **8**, 1445–57.

Cantatore, P., Roberti, M., Rainaldi, G., Gadaleta, M. N., and Saccone, C. (1989). The complete nucleotide sequence, gene organization, and genetic code of the mitochondrial genome of *Paracentrotus lividus*. *J. Biol. Chem.*, **264**, 10965–75.

Capecchi, M. R. and Klein, H. A. (1969). Characterization of three proteins involved in polypeptide chain termination. *Cold Spring Harbor Symp. Quant. Biol.*, **34**, 469–77.

Caron, F. and Meyer, E. (1985). Does *Paramecium primaurelia* use a different genetic code in its macronucleus? *Nature*, **314**, 185–8.

Caskey, C. T. (1980). Peptide chain termination. *Trends Biochem. Sci.*, **5**, 234–7.

Caskey, C. T., Beaudet, A., and Nirenberg, M. (1968). RNA codons and protein synthesis. 15. Dissimilar responses of mammalian and bacterial transfer RNA fractions to messenger RNA codons. *J. Mol. Biol.*, **37**, 99–118.

Caskey, T., Scolnick, E., Tomkins, R., Goldstein, J., and Milman, G. (1969). Peptide chain termination, codon, protein factor, and ribosomal requirements. *Cold Spring Harbor Symp. Quant. Biol.*, **34**, 479–88.

Caspersson, T. O. (1950). *Cell growth and cell function*, p. 185. W. W. Norton and Co., New York.

Cavarelli, J. and Moras, D. (1993). Recognition of tRNAs by aminoacyl-tRNA synthetases. *FASEB J.*, **7**, 79–86.

Cech, T. R. (1986). RNA as an enzyme. *Sci. Amer.*, **255**, 64–75.

Cerretti, D. P., Dean, D., Davis, G. R., Bedwell, D. M., and Nomura, M. (1983). The *spc* ribosomal protein operon of *Escherichia coli*: sequence and cotranscription of the ribosomal protein genes and a protein export gene. *Nucleic Acids Res.*, **11**, 2599–616.

Chambers, I., Frampton, J., Goldfarb, P., Affara, N., McBain, W., and Harrison, P. R. (1986). The structure of the mouse glutathione peroxidase gene: the selenocysteine in the active site is encoded by the 'termination' codon, TGA. *EMBO J.*, **5**, 1221–7.

Chapeville, F., Lipmann, F., von Ehrenstein, G., Weisblum, B., Ray, W. J., Jr., and Benzer, S. (1962). The role of soluble ribonucleic acid in coding for amino acids. *Proc. Natl. Acad. Sci. USA*, **48**, 1086–92.

Chen, S. H., Habib, G., Yang, C.-Y., Gu, Z.-W., Lee, B. R., Weng, S.-A., *et al.* (1987). Apolipoprotein B-48 is the product of a messenger RNA with an organ-specific in-frame stop codon. *J. Biol. Chem.*, **238**, 363–6.

Chevalier, C., Saillard, C., and Bové, J. M. (1990). Organization and nucleotide sequence of the *Spiroplasma citri* genes for ribosomal protein S2, elongation factor Ts, spiralin, phosphofructokinase, pyruvate kinase and an unidentified protein. *J. Bacteriol.*, **172**, 2693–703.

Clark, B. F. C. and Marcker, K. A. (1966). The role of N-formyl-methionyl-sRNA in protein biosynthesis. *J. Mol. Biol.*, **17**, 394–406.

Clark-Walker, G. D., McArthur, C. R., and Spriprakash, K. (1985). Location of transcriptional control signals and transfer RNA sequence in *Torulopsis glabrata* mitochondrial DNA. *EMBO J.*, **4**, 465–73.

Clary, D. O. and Wolstenholme, D. R. (1985). The mitochondrial DNA molecule of *Drosophila yakuba*: nucleotide sequence, gene organization, and genetic code. *J. Mol. Evol.*, **22**, 252–71.

Conklin, P. L., Wilson, R. K., and Hanson, M. R. (1991). Multiple trans-splicing events are required in a plant mitochondrion. *Genes Develop.*, **5**, 1407–15.

Covello, P. S. and Gray, M. W. (1989). RNA editing in plant mitochondria. *Nature*, **341**, 662–6.

Covello, P. S. and Gray, M. W. (1990). Differences in editing at homologous sites in messenger RNAs from angiosperm mitochondria. *Nucleic Acids Res.*, **18**, 5189–96.

Covello, P. S. and Gray, M. W. (1993). On the evolution of RNA editing. *Trends Genet.*, **9**, 265–8.

Cox, E. C. and Yanofsky, C. (1967). Altered base ratios in the DNA of an *Escherichia coli* mutator strain. *Proc. Natl. Acad. Sci. USA*, **58**, 1895–902.

Crick, F. H. C. (1957). Discussion. In *The structure of nucleic acids and their role in protein synthesis*, pp. 25–6. Biochemical Society Symposium no. 14, Cambridge University Press.

Crick, F. H. C. (1958). On protein synthesis. *Symp. Soc. Exp. Biol.*, **12**, 138–63.

Crick, F. H. C. (1963). The recent excitement in the coding problem. *Prog. Nucleic Acid Res.*, **1**, 163–217.

Crick, F. H. C. (1965). Codon–anticodon pairing: the wobble hypothesis. Information Exchange Group No. 7, Nucleic Acids and the Genetic Code. Scientific Memo 14, June 14, 1965.

Crick, F. H. C. (1966*a*). Codon–anticodon pairing: the wobble hypothesis. *J. Mol. Biol.*, **19**, 548–55.

Crick, F. H. C. (1966*b*). The genetic code – yesterday, today, and tomorrow. *Cold Spring Harbor Symp. Quant. Biol.*, **31**, 3–9.

Crick, F. H. C. (1968). The origin of the genetic code. *J. Mol. Biol.*, **38**, 367–79.

Crick, F. H. C., Griffith, J. S., and Orgel, L. E. (1957). Codes without commas. *Proc. Natl. Acad. Sci. USA*, **43**, 416–21.

Crick, F. H. C., Barnett, L., Brenner, S., and Watts-Tobin, R. J. (1961). General nature of the genetic code for protein synthesis. *Nature*, **192**, 1227–32.

Crick, F. H. C., Brenner, S., Klug, A., and Pieczenik, C. (1976). A speculation on the origin of protein synthesis. *Origins of Life*, **7**, 389–97.

Cronin, J. R. and Pizzarello, S. (1983). Amino Acids in meteorites. *Adv. Space Res.*, **3**, 5–18.

de Bruijn, M. H. L. (1983). *Drosophila melanogaster* mitochondrial DNA, a novel organization and genetic code. *Nature*, **304**, 234–41.

Desjardins, P. and Morais, R. (1990). Sequence and gene organization of the chicken mitochondrial genome. A novel gene order in higher vertebrates. *J. Mol. Biol.*, **212**, 599–634.

Dirheimer, G. and Martin, R. P. (1990). Mitochondrial tRNAs: structure, modified nucleosides and codon reading patterns. *J. Chromatogr.*, **45B**, 197–264.

Dounce, A. L. (1952). Duplicating mechanism for peptide chain and nucleic acid synthesis. *Enzymologia*, **15**, 251–8.

Dounce, A. L. (1953). Nucleic acid template hypothesis. *Nature*, **172**, 541.

Dubin, D. T. and HsuChen, C.-C. (1984). Sequence and structure of a methionine transfer RNA from mosquito mitochondria. *Nucleic Acids Res.*, **12**, 4185–9.

Dudler, R., Schmidhauser, C., Parish, R. W., Wettenhall, R. E., and Schmidt, J. (1988). A mycoplasma high-affinity transport system and the *in vitro* invasiveness of mouse sarcoma cells. *EMBO J.*, **7**, 3963–70.

Eigen, M. and Schuster, P. (1978). The hypercycle. A principle of natural self-organization. Part C: The realistic hypercycle. *Naturwiss.*, **65**, 341–69.

Eigen, M., Lindemann, B. F., Tietze, M., Winkler-Oswatitsch, R., Dress, A., and von Haeseler, A. (1989). How old is the genetic code? Statistical geometry of tRNA provides an answer. *Science*, **244**, 673–9.

Elliott, M. S. and Trewyn, R. W. (1984). Inosine biosynthesis in transfer RNA by an enzymatic insertion of hypoxanthine. *J. Biol. Chem.*, **259**, 2407–18.

Epstein, C. J. (1966). Role of the amino acid code and of selection for conformation in the evolution of proteins. *Nature*, **210**, 25–8.

Farabaugh, P.J. (1993). Alternative readings of the genetic code. *Cell*, **74**, 591–6.

Fiers, W., Contreras, R., Duerinck, F., Haegeman, G., Iserentant, D., Merregaert, J., *et al.* (1976). Complete nucleotide sequence of bacteriophage MS2 RNA: primary and secondary structure of the replicase gene. *Nature*, **260**, 500–7.

Filipski, J. (1991). Evolution of DNA sequences. Contribution of mutation bias and selection to the origin of chromosomal compartments. *Adv. Mutagenesis Res.*, **2**, 1–54.

Fitch, W. M. and Upper, K. (1987). The phylogeny of tRNA sequences provides evidence for ambiguity reduction in the origin of the genetic code. *Cold Spring Harbor Symp. Quant. Biol.*, **52**, 759–67.

Forchhammer, K. and Böck, A. (1991). Selenocysteine synthase from *Escherichia coli. J. Biol. Chem.*, **266**, 6324–8.

Forchhammer, K., Leinfelder, W., and Böck, A. (1989). Identification of a novel translation factor necessary for the incorporation of selenocysteine into protein. *Nature*, **342**, 453–6.

Forchhammer, K., Rücknagel, K.-P., and Böck, A. (1990). Purification and biochemical characterization of SELB, a translatic factor involved in selenoprotein synthesis. *J. Biol. Chem.*, **265**, 9346–50.

Förster, C., Ott, G., Forchhammer, K., and Sprinzl, M. (1990). Interaction of a selenocysteine-incorporating tRNA with elongation factor Tu from *E. coli. Nucleic Acids Res.*, **18**, 487–91.

Fox, T. D. (1987). Natural variation in the genetic code. *Annu. Rev. Genet.*, **21**, 67–91.

Fox, T. D. and Leaver, C. J. (1981). The *Zea mays* mitochondrial gene coding cytochrome oxidase subunit II has an intervening sequence and does not contain TGA codons. *Cell*, **26**, 315–23.

Gadaleta, G., Pepe, G., De Candia, G., Quagliariello, C., Sbisa, E., and Saccone, C. (1989). The complete nucleotide sequence of the *Rattus norvegicus* mitochondrial genome: cryptic signals revealed by comparative analysis between vertebrates. *J. Mol. Evol.*, **28**, 497–516.

Gamow, G. (1954). Possible relation between deoxyribonucleic acid and protein structures. *Nature*, **173**, 318.

Garesse, R. (1988). *Drosophila melanogaster* mitochondrial DNA: gene organization and evolutionary considerations. *Genetics*, **118**, 649–83.

Garey, J. R. and Wolstenholme, D. R. (1989). Platyhelminth mitochondrial DNA: evidence for early origin of a $tRNA^{Ser}_{AGN}$ that contains a dihydrouridine arm replacement loop, and serine-specifying AGA and AGG codons. *J. Mol. Evol.*, **28**, 374–87.

Göringer, H. U., Huazi, K. A., Murgola, E. J., and Dahlberg, A. E. (1991). Mutations in 16S rRNA that affect UGA (stop codon)-directed translation termination. *Proc. Natl. Acad. Sci. USA*, **88**, 6603–7.

Grisi, E., Brown, T. A., Waring, R. B., Scazzocchio, C., and Davies, R. W. (1982). Nucleotide sequence of a region of the mitochondrial genome of *Aspergillus nidulans*, including the gene for ATPase subunit 6. *Nucleic Acids Res.*, **10**, 3531–9.

Grivell, L. A. (1986). Deciphering divergent codes. *Nature*, **324**, 109–10.

Grollman, A. P. and Moriya, M. (1993). Mutagenesis by 8-oxoguanine and enemy within. *Trends Genet.*, **9**, 246–9.

Grosjean, H., de Henau, S., and Crothers, D. M. (1978). On the physical basis for ambiguity in genetic coding interactions. *Proc. Natl. Acad. Sci. USA*, **75**, 610–14.

Gualberto, J. M., Lamattina, L., Bonnard, G., Weil, J.-H., and Grienenberger, J.-M. (1989). RNA editing in wheat mitochondria results in the conservation of protein sequences. *Nature*, **341**, 660–2.

Gualberto, J. M., Weil, J.-H., and Grienenberger, J.-M. (1990). Editing of the wheat COXIII transcript: evidence for twelve C to U and one U to C conversions and for sequence similarities around editing sites. *Nucleic Acids Res.*, **18**, 3771–6.

Gupta, R. (1984). *Halobacterium volcanii* tRNAs: identification of 41 tRNAs covering all amino acids, and the sequences of 33 Class I tRNAs. *J. Biol. Chem.*, **259**, 9461–71.

Guthrie, C. and Abelson, J. (1982). Organization and expression of tRNA genes in *Saccharomyces cerevisiae*. In *The molecular biology of the yeast* Saccharomyces: *metabolism and gene expression* (ed. Strathern, J. N., Jones, E., and Broach, J.), pp. 487–528. Cold Spring Harbor Laboratory, New York.

Hanyu, N., Kuchino, Y., Nishimura, S., and Beier, H. (1986). Dramatic events in ciliate evolution: alteration of UAA and UAG termination codons to glutamine codons due to anticodon mutations in two *Tetrahymena* $tRNAs^{Gln}$. *EMBO J.*, **5**, 1307–11.

Hara-Yokoyama, M., Yokoyama, S., Watanabe, T., Watanabe, K., Kitazumi, M., Mitamura, Y., *et al.* (1986). Characteristic anticodon sequences of major tRNA species from an extreme thermophile, *Thermus thermophilus* HB8. *FEBS Lett.*, **202**, 149–52.

Harada, F. and Nishimura, S. (1972). Possible anticodon sequences of $tRNA^{his}$, $tRNA^{asn}$, and $tRNA^{asp}$ from *E. coli* B. Universal presence of nucleoside Q in the first position of the anticodons of these tRNAs. *Biochemistry*, **11**, 301–8.

Harper, D. S. and Jahn, C. L. (1989). Differential use of termination codons in ciliated protozoa. *Proc. Natl. Acad. Sci. USA*, **86**, 3252–6.

Hasegawa, M. and Miyata, T. (1980). On the antisymmetry of the amino acid code table. *Origins of Life*, **10**, 265–70.

Hasegawa, M., Hashimoto, T., Adachi, J., Iwabe, N., and Miyata, T. (1993). Early divergences in the evolution of eukaryotes: ancient divergence of *Entamoeba* that lacks mitochondria revealed by protein sequence data. *J. Mol. Evol.*, **36**, 380–8.

Hatfield, D. and Portugal, F. (1970). Seryl-tRNA in mammalian tissues: chromatographic differences in brain and liver and a specific response to the codon UGA. *Proc. Natl. Acad. Sci. USA*, **67**, 1200–6.

Hatfield, D., Choi, I. S., Mischke, S., and Owens, L. D. (1992). Selenocysteyl-tRNAs recognize UGA in *Beta vulgaris*, a higher plant, and in *Gliocladium virens*, a filamentous fungus. *Biochem. Biophys. Res. Commun.*, **184**, 254–9.

Haurowitz, F. (1950). *Chemistry and biology of proteins*, 374, Academic Press, New York.

Haurowitz, F. and Crampton, C. F. (1952). The role of the nucleus in protein synthesis. *Exp. Cell Res.*, **Suppl. 2**, 45–57.

Heckman, J. E., Sarnoff, J., Alzner-DeWeerd, B., Yin, S., and RajBhandary, U. L. (1980). Novel features in the genetic code and codon reading patterns in *Neurospora crassa* mitochondria based on sequences of six mitochondrial tRNAs. *Proc. Natl. Acad. Sci. USA*, **77**, 3159–63.

Heider, J., Leinfelder, W., and Böck, A. (1989). Occurrence and functional compatibility within Enterobacteriaceae of tRNA species which inserts selenocysteine into protein. *Nucleic Acids Res.*, **17**, 2529–40.

Helftenbein, E. (1985). Nucleotide sequence of a macronuclear DNA molecule coding for alpha-tubulin from the ciliate *Stylonichia lemnae*. Special codon usage: TAA is not a translation termination codon. *Nucleic Acids Res.*, **13**, 415–33.

Helftenbein, E. and Müller, E. (1988). Both alpha-tubulin genes are transcriptionally active in *Stylonichia lemnae. Curr. Genet.*, **13**, 425–32.

Hendriks, L., Van Broeckhoven, C., Vandenberghe, A., Van De Peer, Y., and DeWachter, R. (1988). Primary and secondary structure of the 18S ribosomal RNA of the bird spider *Eurypelma californica* and evolutionary relationship among eukaryotic phyla. *Eur. J. Biochem.*, 177, 15–20.

Hensgens, L. A. M., Grivell, L. A., Borst, P., and Bos, J. L. (1979). Nucleotide sequence of the mitochondrial structural gene for subunit 9 of yeast ATPase complex. *Proc. Natl. Acad. Sci. USA*, **76**, 1663–7.

Hensgens, L. A. M., Brakenhoff, J., De Vries, B. F., Sloof, P., Tromp, M. C., Van Boon, J. H., and Benne, R. (1984). The sequence of the gene for cytochrome c oxidase subunit I, a frameshift containing gene for cytochrome c oxidase subunit II and seven unassigned reading frames in *Trypanosoma brucei* mitochondrial maxi-circle DNA. *Nucleic Acids Res.*, **12**, 7327–44.

Herrick, G., Hunter, D., Williams, K., and Kotter, K. (1987). Alternate processing during development of a macronuclear chromosome family in *Oxytricha fallax. Genes Develop.*, **1**, 1047–58.

Hiesel, R. and Brennicke, A. (1983). Cytochrome oxidase subunit II gene in mitochondria of *Oenothera* has no intron. *EMBO J.*, **2**, 2173–8.

Hiesel, R., Wissinger, B., Schuster, W., and Brennicke, A. (1989). RNA editing in plant mitochondria. *Science*, **246**, 1632–4.

Hill, K. E., Lloyd, R. S., Yang, J.-G., Read, R., and Burk, R. F. (1991). The cDNA for rat selenoprotein P contains 10 TGA codons in the open reading frame. *J. Biol. Chem.*, **266**, 10050–3.

Himeno, H., Masaki, H., Ohta, T., Kumagai, I., Miura, K.-I., and Watanabe, K. (1987). Unusual genetic codes and a novel genome structure for $tRNA^{Ser}_{AGY}$ in starfish mitochondrial DNA. *Gene*, **56**, 219–30.

Hiratsuka, J., Shimada, H., Whittier, R., Ishibashi, T., Sakamoto, M., Mori, M., *et al.* (1989). The complete sequence of the rice (*Oryza sativa*) chloroplast genome: intermolecular recombination between distinct tRNA genes accounts for a major plastid DNA inversion during the evolution of the cereals. *Mol. Gen. Genet.*, **217**, 185–94.

Hoagland, M. B. (1960). The relationship of nucleic acid and protein synthesis as revealed by studies in cell-free systems. In *The nucleic acids*, Vol. II (ed. Chargaff, E., and Davidson, J. N.), pp. 349–408, Academic Press, New York.

Hoch, B., Maier, R. M., Appel, K., Igloi, G. L., and Kossel, H. (1991). Editing of a chloroplast mRNA by creation of an initiation codon. *Nature*, **353**, 178–80.

Hofmann, J., Schumann, G., Borschet, G., Gösseringer, R., Bach, M., Bertling, W. M., *et al.* (1991). Transfer RNA genes from *Dictyostelium discoideum* are frequently associated with repetitive elements and contain consensus boxes in their 5′ and 3′-flanking regions. *J. Mol. Biol.*, **222**, 537–52.

Hoffmann, R. J., Boore, J. L., and Brown, W. M. (1992). A novel mitochondrial genome organization for the blue mussel, *Mytilus edulis*. *Genetics*, **131**, 397–412.
Holley, R. W., Apgar, J., Everett, G. A., Madison, J. T., Marquisse, M., Merrill, S. H., Penswick, J. R., and Zamir, A. (1965). Structure of a ribonucleic acid. *Science*, **147**, 1462–5.
Holmquist, G. P. (1989). Evolution of chromosome bands: molecular ecology of non-coding DNA. *J. Mol. Evol.*, **28**, 469–86.
Holmquist, W. R. (1978). Evolution of compositional non-randomness in proteins. *J. Mol. Evol.*, **11**, 349.
Holmquist, R., Jukes, T. H., and Pangburn, S. (1973). Evolution of transfer RNA. *J. Mol. Biol.*, **78**, 91–116.
Hori, H. and Osawa, S. (1979). Evolutionary change in 5S RNA secondary structure and a phylogenic tree of 54 5S RNA species. *Proc. Natl. Acad. Sci. USA*, **76**, 381–5.
Hori, H. and Osawa, S. (1987). Origin and evolution of organisms as deduced from 5S ribosomal RNA sequences. *Mol. Biol. Evol.*, **4**, 445–72.
Horowitz, S. and Gorovsky, M. A. (1985). An unusual genetic code in nuclear genes of *Tetrahymena*. *Proc. Natl. Acad. Sci. USA*, **82**, 2452–5.
Hou, Y.-M. and Schimmel, P. (1988). A simple structural feature is a major determinant of the identity of a transfer RNA. *Nature*, **333**, 140–5.
HsuChen, C.-C. and Dubin, D. T. (1984). A cluster of four transfer RNA genes in mosquito mitochondrial DNA. *Biochem. Intern.*, **8**, 385–91.
HsuChen, C.-C., Cleaves, G. R., and Dubin, D. T. (1983). A major lysine tRNA with a CUU anticodon in insect mitochondria. *Nucleic Acids Res.*, **11**, 8659–62.
Hudspeth, M. E. S., Ainley, W. M., Shumard, D. S., Butow, R. A., and Grossman, L. I. (1982). Location and structure of the *var1* gene on yeast mitochondrial DNA: nucleotide sequence of the 40.0 allele. *Cell*, **30**, 617–26.
Huttenhofer, A. and Noller, H. F. (1992). Hydroxyl radical cleavage of tRNA in the ribosomal P-site. *Proc. Natl. Acad. Sci. USA*, **89**, 7851–5.
Ikeda, R., Ohama, T., Muto, A., and Osawa, S. (1990*a*). Nucleotide sequences of nine tRNA genes from *Micrococcus luteus*. *Nucleic Acids Res.*, **18**, 7154.
Ikeda, R., Ohama, T., Muto, A., and Osawa, S. (1990*b*). Nucleotide sequences of two tRNA gene clusters from *Micrococcus luteus*. *Nucleic Acids Res.*, **18**, 7155.
Ikemura, T. (1981*a*). Correlation between the abundance of *Escherichia coli* transfer RNAs and the occurrence of the respective codons in its protein genes. *J. Mol. Biol.*, **146**, 1–21.
Ikemura, T. (1981*b*). Correlation between the abundance of *Escherichia coli* transfer RNAs and the occurrence of the respective codons in its protein genes: a proposal for a synonymous codon choice that is optimal for the *E. coli* translational system. *J. Mol. Biol.*, **151**, 389–409.
Ikemura, T. (1982). Correlation between the abundance of yeast transfer RNAs and the occurrence of the respective codons in protein genes. *J. Mol. Biol.*, **158**, 573–97.
Ikemura, T. and Aota, S.-I. (1988). Global variation in G+C content along vertebrate genome DNA. Possible correlation with chromosome band structure. *J. Mol. Evol.*, **203**, 1–13.
Illangasekare, M., Sanchez, G., Nickles, T., and Yarus, M. (1995). Aminoacyl-RNA synthesis catalyzed by an RNA. *Science*, **267**, 643–7.
Inagaki, Y., Bessho, Y., and Osawa, S. (1993). Lack of peptide-release activity responding to codon UGA in *Mycoplasma capricolum*. *Nucleic Acids Res.*, **21**, 1335–8.
Inamine, J. M., Denny, T. P., Loechel, S., Schaper, U., Huang, C.-H., Bott, K. F., and Hu, P.-C. (1988). Nucleotide sequence of the P1 attachment-protein gene of *Mycoplasma pneumoniae*. *Gene*, **64**, 217–29.

Inamine, J. M., Ho, K., Loechel, S., and Hu, P. (1990). Evidence that UGA is read as tryptophan rather than stop by *Mycoplasma pneumoniae*, *Mycoplasma genitalium* and *Mycoplasma gallisepticum*. *J. Bacteriol.*, **172**, 504–6.

Iwabe, N., Kuma, K., Hasegawa, M., Osawa, S., and Miyata, T. (1989). Evolutionary relationship of archaebacteria, eubacteria, and eukaryotes inferred from phylogenetic trees of duplicated genes. *Proc. Natl. Acad. Sci. USA*, **86**, 9355–9.

Jacobs, H. T., Elliott, D. J., Math, V. B., and Farquharson, A. (1988). Nucleotide sequence and gene organization of sea-urchin mitochondrial DNA. *J. Mol. Biol.*, **202**, 185–217.

Jacobs, H. T., Asakawa, S., Araki, T., Miura, K., Smith, M., and Watanabe, K. (1989). Conserved tRNA gene cluster in starfish mitochondrial DNA. *Curr. Genet.*, **15**, 193–206.

Janke, A. and Pääbo, S. (1993). Editing of a tRNA anticodon in marsupial mitochondria changes its codon recognition. *Nucleic Acids Res.*, **21**, 1523–5.

Jukes, T. H. (1966). *Molecules and evolution*, p. 284. Columbia University Press, New York.

Jukes, T. H. (1973*a*). Possibilities for the evolution of the genetic code from a preceding form. *Nature*, **246**, 22–6.

Jukes, T. H. (1973*b*). Arginine as an evolutionary intruder into protein synthesis. *Biochem. Biophys. Res. Commun.*, **53**, 709–14.

Jukes, T. H. (1977*a*). How many anticodons? *Science*, **198**, 319–20.

Jukes, T. H. (1977*b*). The amino acid code. In *Comprehensive biochemistry, Vol. 24 – Biological information transfer* (ed. Neuberger, A.), pp. 235–93. Elsevier, Amsterdam.

Jukes, T. H. (1981). Amino acid codes in mitochondria as possible clues to primitive codes. *J. Mol. Evol.*, **18**, 15–17.

Jukes, T. H. (1983*a*). Evolution of the amino acid code: inferences from mitochondrial codes. *J. Mol. Evol.*, **19**, 219–25.

Jukes, T. H. (1983*b*). Evolution of the amino acid code. In *Evolution of genes and proteins* (ed. Nei, M. and Koehn, R.), pp. 191–207. Sinaauer Associates Inc., Sunderland, MA.

Jukes, T. H. (1985). A change in the genetic code in *Mycoplasma capricolum*. *J. Mol. Evol.*, **22**, 361–2.

Jukes, T. H. (1990). Genetic code 1990. Outlook. *Experientia*, **46**, 1149–57.

Jukes, T. H. and Bhushan, V. (1986). Silent nucleotide substitutions and G + C content of some mitochondrial and bacterial genes. *J. Mol. Evol.*, **24**, 39–44.

Jukes, T. H. and Osawa, S. (1990). The genetic code in mitochondria and chloroplast. *Experientia*, **46**, 1117–26.

Jukes, T. H. and Osawa, S. (1991). Recent evidence for evolution of the genetic code. In *Evolution of life: fossils, molecules and culture* (ed. Osawa, S. and Honjo, T.), pp. 79–95. Springer-Verlag, Tokyo.

Jukes, T. H. and Osawa, S. (1993). Evolutionary changes in the genetic code. *Comp. Biochem. Physiol.*, **106B**, 489–94.

Jukes, T. H., Holmquist, R., and Moise, H. (1975). Amino acid composition of proteins: selection against the genetic code. *Science*, **189**, 50–1.

Jukes, T. H., Osawa, S., and Muto, A. (1987). Divergence and directional mutation pressures. *Nature*, **325**, 668.

Jukes, T. H., Bessho, Y., Ohama, T., and Osawa, S. (1991). Release factors and genetic code. *Nature*, **352**, 575.

Kagawa, Y., Nojima, H., Nukiwa, N., Ishizuka, M., Nakajima, T., Yasuhara, T., *et al.* (1984). High guanine plus cytosine content in the third letter of codons of extreme thermophile. DNA sequence of the isopropylmalate dehydrogenase of *Thermus thermophilus*. *J. Biol. Chem.*, **259**, 2956–60.

Kano, A., Andachi, Y., Ohama, T., and Osawa, S. (1991). Novel anticodon composition of transfer RNAs in *Micrococcus luteus*, a bacterium with a high genomic G + C-content: correlation with codon usage. *J. Mol. Biol.*, **221**, 387–401.

Kano, A., Ohama, T., Abe, R., and Osawa, S. (1993). Unassigned or nonsense codons in *Micrococcus luteus*. *J. Mol. Biol.*, **230**, 51–6.

Karlovsky, P. and Fartmann, B. (1992). Genetic code and phylogenetic origin of oomycetous mitochondria. *J. Mol. Evol.*, **34**, 254–8.

Kawaguchi, Y., Honda, H., Taniguchi-Morimura, J., and Iwasaki, S. (1989). The codon CUG is read as serine in an asporogenic yeast *Candida cylindracea*. *Nature*, **341**, 164–6.

Khorana, H. G., Büchi, H., Ghosh, H., Gupta, N., Jacob, T. M., Kössel, H., *et al.* (1966). Polynucleotide synthesis and the genetic code. *Cold Spring Harbor Symp. Quant. Biol.*, **31**, 39–49.

Kim, S. H., Suddath, F. L., Quigley, G. J., McPherson, A., Sussman, J. L., Wang, A. H. J., *et al.* (1974*a*). Three-dimensional tertiary structure of yeast phenylalanine transfer RNA. *Science*, **185**, 435–40.

Kim, S. H., Sussman, J. L., Suddath, F. L., Quigley, G. J., McPherson, A., Wang, A. H. J., *et al.* (1974*b*). The general structure of transfer RNA molecule. *Proc. Natl. Acad. Sci. USA*, **71**, 4970–4.

Kimura, M. (1983). *The neutral theory of molecular evolution*, p. 367. Cambridge University Press, Cambridge.

King, J. L. and Jukes, T. H. (1969). Non-Darwinian evolution. *Science*, **164**, 788–98.

Knoop, V., Schuster, W., Wissinger, B., and Brennicke, A. (1991). Trans-splicing integrates an exon of 22 nucleotides into the nad 5 mRNA in higher plant mitochondria. *EMBO J.*, **10**, 3483–93.

Köchel, H. G., Lazarus, C. M., Basak, N., and Küntzel, H. (1981). Mitochondrial tRNA gene clusters in *Aspergillus nidulans*: organizations and nucleotide sequence. *Cell*, **23**, 625–33.

Komine, Y., Adachi, T., Inokuchi, H., and Ozeki, H. (1990). Genomic organization and physical mapping of the transfer RNA genes in *Escherichia coli* K12. *J. Mol. Biol.*, **212**, 579–98.

Kondo, K., Sone, H., Yoshida, H., Toida, T., Kanatani, K., Hong, Y.-M., *et al.* (1990). Cloning and sequence analysis of the arginine deaminase gene from *Mycoplasma arginini*. *Mol. Gen. Genet.*, **221**, 81–6.

Kuchino, Y., Hanyu, N., Tashiro, F., and Nishimura, S. (1985). *Tetrahymena thermophila* glutamine tRNA and its gene that corresponds to UAA termination codon. *Proc. Natl. Acad. Sci. USA*, **82**, 4758–62.

Kudla, J., Igloi, G. L., Metzlaff, M., Hagemann, R., and Kössel, H. (1992). RNA editing in tobacco chloroplasts leads the formation of a translatable *psbL* mRNA by a C to U substitution within the initiation codon. *EMBO J.*, **11**, 1099–103.

Kumazawa, Y., Yokogawa, T., Hasegawa, E., Miura, K., and Watanabe, K. (1989). The aminoacylation of structurally variant phenylalanine tRNAs from mitochondria and various nonmitochondrial sources by bovine mitochondrial phenylalanyl-tRNA synthetase. *J. Biol. Chem.*, **264**, 13005–11.

Kumazawa, Y., Himeno, H., Miura, K., and Watanabe, K. (1991). Unilateral aminoacylation specificity between bovine mitochondria and eubacteria. *J. Biochem.*, **109**, 421–7.

Kurland, C. G. (1992). Evolution of mitochondrial genomes and the genetic code. *BioEssays*, **14**, 347–52.

Lagerkvist, U. (1978). 'Two out of three': an alternative method for codon reading. *Proc. Natl. Acad. Sci. USA*, **75**, 1759–62.

Lagerkvist, U. (1981). Unorthodox codon reading and the evolution of the genetic code. *Cell*, **23**, 305–6.

Lamattina, L. and Grienenberger, J. M. (1991). RNA editing of the transcript coding for subunit 4 of NADH dehydrogenase in wheat mitochondria: uneven distribution of the editing sites among four exons. *Nucleic Acids Res.*, **19**, 3275–82.

Lee, B. J., Rajagopalan, M., Kim, Y. S., Kwang, H., You, K., Jacobson, K. B., and Hatfield, D. (1990). Selenocysteine $tRNA^{[Ser]Sec}$ gene is ubiquitous within the animal kingdom. *Mol. Cell. Biol.*, **10**, 1940–9.

Lee, C. C., Timms, K. M., Trotman, C. N. A., and Tate, W. P. (1987). Isolation of a rat mitochondrial release factor: accommodation of the changed genetic code for termination. *J. Biol. Chem.*, **262**, 3548–52.

Leinfelder, W., Zehelein, E., Mandrand-Berthelot, M. A., and Böck, A. (1988). Gene for a novel tRNA species that accepts L-serine and co-translationally inserts selenocysteine. *Nature*, **331**, 723–5.

Leinfelder, W., Forchhammer, K., Veprek, B., Zehelein, E., and Böck, A. (1990). *In vitro* synthesis of selenocysteinyl-$tRNA_{UCA}$ from seryl-$tRNA_{UCA}$: involvement and characterization of *sel*D gene product. *Proc. Natl. Acad. Sci. USA*, **87**, 543–7.

Lewin, B. (1974). *Gene expression*. Vol. 1, (pp. 30–7). John Wiley and Sons, London.

Lewin, B. (1990). *Genes IV*, (p. 148). Oxford University Press, Oxford.

Li, M. and Tzagoloff, A. (1979). Assembly of the mitochondrial membrane system: sequences of yeast mitochondrial valine and an unusual threonine tRNA gene. *Cell*, **18**, 47–53.

Li, S., Pelka, H., and Schulman, L. H. (1993). The anticodon and discriminator base are important for aminoacylation of *Escherichia coli* $tRNA^{Asn}$. *J. Biol. Chem.*, **268**, 18335–9.

Liu, H. and Beckenbach, A. T. (1992). Evolution of the mitochondrial cytochrome oxidase II gene among 10 orders of insects. *Mol. Phylogen. Evol.*, **1**, 41–52.

Lonergan, K. M. and Gray, M. W. (1993). Editing of transfer RNAs in *Acanthamoeba castellanii* mitochondria. *Science*, **259**, 812–15.

Lustig, F., Borén, T., Claesson, C., Simonsson, C., Barciszewska, M., and Lagerkvist, U. (1993). The nucleotide in position 32 of the tRNA anticodon loop determines ability of anticodon UCC to discriminate among glycine codons. *Proc. Natl. Acad. Sci. USA*, **90**, 3343–7.

Macino, G. and Tzagoloff, A. (1979). Assembly of the mitochondrial membrane system – partial sequence of a mitochondrial ATPase gene in *Saccharomyces cerevisiae. Proc. Natl. Acad. Sci. USA*, **76**, 131–5.

Macino, G., Coruzzi, G., Nobrega, F. G., Li, M., and Tzagoloff, A. (1979). Use of the UGA terminator as a tryptophan codon in yeast mitochondria. *Proc. Natl. Acad. Sci. USA*, **76**, 3784–5.

Maizels, N. and Weiner, A. M. (1987). Peptide-specific ribosomes, genomic tags, and the origin of the genetic code. *Cold Spring Harbor Symp. Quant. Biol.*, **52**, 743–57.

Maki, H. and Sekiguchi, M. (1992). *Mut*T protein specifically hydrolyses a potent mutagenic substrate for DNA synthesis. *Nature*, **355**, 273–5.

Marcker, K. and Sanger, F. (1964). *N*-Formyl-methionyl-sRNA. *J. Mol. Biol.*, **8**, 835–40.

Maréchal, L., Guillemaut, P., Grienenberger, J.-M., Jeannin, G., and Weil, J.-H. (1985). Sequence and codon recognition of bean mitochondria and chloroplast $tRNAs^{Trp}$: evidence for a high degree of homology. *Nucleic Acids Res.*, **13**, 4411–6.

Maréchal, L., Runeberg-Roos, P., Grinenberger, J. M., Colin, J., Weil, J.-H., Lejeune, B., *et al.* (1987). Homology in the region containing a $tRNA^{Trp}$ gene and a (complete or partial) $tRNA^{Pro}$ gene in wheat mitochondrial and chloroplast genomes. *Curr. Genet.*, **12**, 91–8.

Maréchal-Drouard, L., Guillemaut, P., Cosset, A., Arbogast, M., Weber, F., Weil, J., and Dietrich, A. (1990). Transfer RNAs of potato (*Solanum tuberosum*) mitochondria have different genetic origins. *Nucleic Acids Res.*, **18**, 3689–96.

Martin, R. P., Schneller, J.-M., Stahl, A. J. C., and Dirheimer, G. (1979). Import of nuclear DNA coded lysine-accepting tRNA (anticodon C-U-U) into yeast mitochondria. *Biochemistry*, **18**, 4600–5.

Martin, R. P., Sibler, A.-P., and Dirheimer, G. (1982). The primary structure of three yeast mitochondrial serine tRNA isoacceptors. *Biochimie*, **64**, 1073–9.

Martin, R. P., Sibler, A.-P., Gehrke, C. W., Kuo, K., Edmonds, C. G., McCloskey, J. A., and Dirheimer, G. (1990). 5-[[(carboxymethyl)amino]methyl]uridine is found in the anticodon of yeast mitochondrial tRNAs recognizing two-codon families ending in a purine. *Biochemistry*, **29**, 956–9.

Matthaei, J. H., Vogt, H. P., Neth, H. R., Schöch, G., Kübler, H., Amelunxen, F., Sander, G., and Parmeggiani, A. (1966). Specific interactions of ribosomes in decoding. *Cold Spring Harbor Symp. Quant. Biol.*, **31**, 25–38.

McClain, W. H. (1993). Rules that govern tRNA identity in protein synthesis. *J. Mol. Biol.*, **234**, 257–80.

Meyer, F. M., Schmidt, H. J., Plümper, E., Hasilik, A., Mersmann, G., Meyer, H.E., *et al.* (1991). UGA is translated as cysteine in pheromone 3 of *Euplotes octocarinatus. Proc. Natl. Acad. Sci. USA*, **88**, 3758–61.

Miller, S. L. (1987). Which organic compounds could have occurred on the prebiotic earth? *Cold Spring Harbor Symp. Quant. Biol.*, **52**, 17–27.

Moazed, D. and Noller, H. F. (1991). Sites of interaction of the CCA end of peptidyl-tRNA with 23S rRNA. *Proc. Natl. Acad. Sci. USA*, **88**, 3725–8.

Moriya, J., Yokogawa, T., Wakita, K., Ueda, T., Nishikawa, K., Crain, P. F., *et al.* (1994). A novel modified nucleoside found at the first position of the anticodon of methionine tRNA from bovine liver mitochondria. *Biochemistry*, **33**, 2234–9.

Munz, P., Leupold, U., Agris, P., and Kohli, J. (1981). *In vivo* decoding rules in *Schizosaccharomyces pombe* are at variance with *in vitro* data. *Nature*, **294**, 187–8.

Muramatsu, T., Nishikawa, K., Nemoto, F., Kuchino, Y., Nishimura, S., Miyazawa, T., and Yokoyama, S. (1988*a*). Codon and amino acid specificities of a transfer RNA are both converted by a single post-transcriptional modification. *Nature*, **336**, 179–81.

Muramatsu, T., Yokoyama, S., Horie, N., Matsuda, A., Ueda, T., Yamaizumi, Z., *et al.* (1988*b*). A novel lysine-substituted nucleoside in the first position of the anticodon of minor isoleucine tRNA from *Escherichia coli. J. Biol. Chem.*, **263**, 9261–7.

Murao, K. and Ishikura, H. (1978). A new uridine derivative located in the anticodon of $tRNA^{Gly}$ from *Bacillus subtilis. Nucleic Acids Res.*, **5**, S333–6.

Muto, A. and Osawa, S. (1987). The guanine and cytosine content of genomic DNA and bacterial evolution. *Proc. Natl. Acad. Sci. USA*, **84**, 166–9.

Muto, A., Andachi, Y., Yuzawa, H., Yamao, F., and Osawa, S. (1990). The organization and evolution of transfer RNA genes of *Mycoplasma capricolum. Nucleic Acids Res.*, **18**, 5037–43.

Muto, A., Ohama, T., Andachi, Y., Yamao, F., Tanaka, R., and Osawa, S. (1991). Evolution of codons and anticodons in eubacteria. In *New aspects in population genetics and molecular evolution*, (ed. Kimura, M. and Takahata, N.), pp. 179–93. Japan Scientific Society Press/Springer-Verlag, Tokyo.

Nagel, G. M. and Doolittle, R. F. (1991). Evolution and relatedness in two aminoacyl-tRNA synthetase families. *Proc. Natl. Acad. Sci. USA*, **88**, 8121–5.

Nellen, W. and Gallwitz, D. (1982). Actin genes and actin messenger RNA in *Acanthamoeba castellanii*: nucleotide sequence of the split actin gene I. *J. Mol. Biol.*, **159**, 1–18.

Netzker, R., Köchel, H. G., Basak, N., and Küntzel, H. (1982). Nucleotide sequence of *Aspergillus nidulans* mitochondrial genes coding for ATPase subunit 6, cytochrome oxidase subunit 3, seven unidentified proteins, four tRNAs and L-rRNA. *Nucleic Acids Res.*, **10**, 4783–94.

Nghiem, Y., Cabrera, M., Cupples, C. G., and Miller, J. H. (1988). The *mut*Y gene: a mutator locus in *Escherichia coli* that generates G·C→T·A transversions. *Proc. Natl. Acad. Sci. USA*, **85**, 2709–13.

Nirenberg, M. and Leder, P. (1964). RNA codewords and protein synthesis. *Science*, **145**, 1399–1407.

Nirenberg, M. W. and Matthaei, J. H. (1961). The dependence of cell-free protein synthesis in *E. coli* upon naturally occurring or synthetic polyribonucleotides. *Proc. Natl. Acad. Sci. USA*, **47**, 1588–602.

Nirenberg, M., Caskey, T., Marshal, R., Brimacombe, R., Kellogg, D., Doctor, B., *et al.* (1966). The RNA code and protein synthesis. *Cold Spring Harbor Symp. Quant. Biol.*, **31**, 11–24.

Nishimura, S. (1979). Modified nucleosides in tRNA. In *Transfer RNA: structure, properties, and recognition* (ed. Schimmel, P. R., Söll, D., and Abelson, J. M.), pp. 59–79. Cold Spring Harbor Laboratory, New York.

Nishimura, S. (1983). Structure, biosynthesis and function of queuosine in tRNA. *Prog. Nucleic Acid Res. Mol. Biol.*, **28**, 49–73.

Nishimura, S., Jones, D. S., and Khorana, H. G. (1965). The *in vitro* synthesis of a co-polypeptide containing two amino acids in alternating sequence dependent upon a DNA-like polymer containing two nucleotides in alternating sequence. *J. Mol. Biol.*, **13**, 302–24.

Nitta, I., Ueda, T., and Watanabe, K. (1994). Template-dependent peptide formation on ribosomes catalyzed by pyridine. *J. Biochem.*, **115**, 803–7.

Noller, H. F., Hoffarth, V., and Zimniak, L. (1992). Unusual resistance of peptidyl transferase to protein extraction procedures. *Science*, **256**, 1416–19.

Nomoto, M., Imai, N., Saiga, H., Matsui, T., and Mita, T. (1987). Characterization of two types of histone H2B genes from macronuclei of *Tetrahymena thermophila*. *Nucleic Acids Res.*, **15**, 5681–98.

Normanly, J. and Abelson, J. (1989). tRNA identity. *Annu. Rev. Biochem.*, **58**, 1029–49.

Nowak, C. and Kuck, U. (1990). RNA editing of the mitochondrial *atp*9 transcript from wheat. *Nucleic Acids Res.*, **18**, 7164.

Nureki, O., Niimi, T., Muramatsu, T., Kanno, H., Kohno, T., Florentz, C., *et al.* (1994). Molecular recognition of the identity-determinant set of isoleucine transfer RNA from *Escherichia coli*. *J. Mol. Biol.*, **236**, 710–24.

Oba, T., Andachi, Y., Muto, A., and Osawa, S. (1991*a*). CGG, unassigned or nonsense codon: Occurrence in *Mycoplasma capricolum*. *Proc. Natl. Acad. Sci. USA*, **88**, 921–5.

Oba, T., Andachi, Y., Muto, A., and Osawa, S. (1991*b*). Translation *in vitro* of codon UGA as tryptophan in *Mycoplasma capricolum*. *Biochimie*, **73**, 1109–12.

Ohama, T., Yamao, F., Muto, A., and Osawa, S. (1987). Organization and codon usage of the streptomycin operon in *Micrococcus luteus*, a bacterium with a high genomic G+C content. *J. Bacteriol.*, **169**, 4770–7.

Ohama, T., Muto, A., and Osawa, S. (1989). Spectinomycin operon of *Micrococcus luteus*: evolutionary implications of organization and codon usage. *J. Mol. Evol.*, **29**, 381–95.

Ohama, T., Muto, A., and Osawa, S. (1990*a*). Role of GC-biased mutation pressure on synonymous codon choice in *Micrococcus luteus*, a bacterium with a high genomic GC-content. *Nucleic Acids Res.*, **18**, 1565–9.

Ohama, T., Osawa, S., Watanabe, K., and Jukes, T. H. (1990*b*). Evolution of the mitochondrial genetic code IV. AAA as an asparagine codon in some animal mitochondria. *J. Mol. Evol.*, **30**, 329–32.

Ohama, T., Suzuki, T., Mori, M., Osawa, S., Ueda, T., Watanabe, K., and Nakase, T. (1993). Non-universal decoding of the leucine codon CUG in several *Candida* species. *Nucleic Acids Res.*, **21**, 4039–45.

Ohama, T., Yang, D. C.-H., and Hatfield, D. L. (1994). Identity elements within mammalian $tRNA^{(Ser)Sec}$ and $tRNA^{Ser}$ for seryl-tRNA synthetase recognition. *Arch. Biochem. Biophys.*, **315**, 293–301.

Ohashi, Z., Saneyoshi, M., Harada, H., Hara, H., and Nishimura, S. (1970). Presumed anticodon structure of glutamic acid tRNA from *E. coli*: a possible location of a 2-thiouridine derivative in the first position of the anticodon. *Biochem. Biophys. Res. Commun.*, **40**, 866–72.

Ohkubo, S., Muto, A., Kawauchi, Y., Yamao, F., and Osawa, S. (1987). The ribosomal protein gene cluster of *Mycoplasma capricolum*. *Mol. Gen. Genet.*, **210**, 314–22.

Ohta, T. and Kimura, M. (1971). Amino acid composition of proteins as a product of molecular evolution. *Science*, **174**, 150–3.

Ohyama, K., Fukuzawa, H., Kohchi, T., Shirai, H., Sano, T., Sano, S., *et al.* (1986). Chloroplast gene organization deduced from complete sequence of liverwort *Marchantia polymorpha* chloroplast DNA. *Nature*, **322**, 572–4.

Ohyama, K., Fukuzawa, H., Kohchi, T., Sano, T., Sano, S., Shirai, H., *et al.* (1988). Structure and organization of *Marchantia polymorpha* chloroplast genome. I. Cloning and gene identification. *J. Mol. Biol.*, **203**, 281–98.

Ohyama, K., Ogura, Y., Oda, K., Yamato, K., Ohta, E., Nakamura, Y., *et al.* (1991). Evolution of organellar genomes. In *Evolution of life: fossils, molecules, and culture* (ed. Osawa, S. and Honjo, T.), pp. 187–98. Springer-Verlag, Tokyo.

Okada, Y., Terzaghi, E., Streisinger, G., Emrich, J., Inouye, M., and Tsugita, A. (1966). A frame-shift mutation involving the addition of two base pairs in the lysozyme gene of phage T4. *Proc. Natl. Acad. Sci. USA*, **56**, 1692–8.

Okimoto, R., Macfarlane, J. L., Clary, D. O., and Wolstenholme, D. R. (1992). The mitochondrial genomes of two nematodes, *Caenorhabditis elegans* and *Ascaris suum*. *Genetics*, **130**, 471–98.

Orgel, L. E. and Crick, F. H. C. (1993). Anticipating an RNA world. Some past speculations on the origin of life: where are they today? *FASEB J.*, **7**, 238–9.

Osawa, S. and Jukes, T. H. (1988). Evolution of the genetic code as affected by anticodon content. *Trends Genet.*, **4**, 191–8.

Osawa, S. and Jukes, T. H. (1989). Codon reassignment (codon capture) in evolution. *J. Mol. Evol.*, **28**, 271–8.

Osawa, S., Jukes, T. H., Muto, A., Yamao, F., Ohama, T., and Andachi, Y. (1987). Role of directional mutation pressure in the evolution of the eubacterial genetic code. *Cold Spring Harbor Symp. Quant. Biol.*, **52**, 777–89.

Osawa, S., Ohama, T., Yamao, F., Muto, A., Jukes, T. H., Ozeki, H., and Umesono, K. (1988). Directional mutation pressure and transfer RNA in choice of the third nucleotide of synonymous two-codon sets. *Proc. Natl. Acad. Sci. USA*, **85**, 1124–8.

Osawa, S., Ohama, T., Jukes, T. H., and Watanabe, K. (1989*a*). Evolution of the mitochondrial genetic code. I. Origin of AGR serine and stop codons in metazoan mitochondria. *J. Mol. Evol.*, **29**, 202–7.

Osawa, S., Ohama, T., Jukes, T. H., Watanabe, K., and Yokoyama, S. (1989*b*). Evolution of the mitochondrial genetic code. II. Reassignment of codon AUA from isoleucine to methionine. *J. Mol. Evol.*, **29**, 373–80.

Osawa, S., Collins, D., Ohama, T., Jukes, T. H., and Watanabe, K. (1990*a*). Evolution of the mitochondrial genetic code. III. Reassignment of CUN codons from leucine to threonine during evolution of yeast mitochondria. *J. Mol. Evol.*, **30**, 322–8.

Osawa, S., Muto, A., Jukes, T. H., and Ohama, T. (1990*b*). Evolutionary changes in the genetic code. *Proc. Roy. Soc. London Ser. B*, **241**, 19–28.

Osawa, S., Muto, A., Ohama, T., Andachi, Y., Tanaka, R., and Yamao, F. (1990*c*). Prokaryotic genetic code. *Experientia*, **46**, 1097–106.

Osawa, S., Jukes, T. H., Watanabe, K., and Muto, A. (1992). Recent evidence for evolution of the genetic code. *Microbiological Rev.*, **56**, 229–64.

Ozeki, H., Ohyama, K., Inokuchi, H., Fukuzawa, H., Kohchi, T., Sato, T., *et al.* (1987). Genetic system of chloroplasts. *Cold Spring Harbor Symp. Quant. Biol.*, **52**, 791–804.

Pape, L. K. and Tzagoloff, A. (1985). Cloning and characterization of the gene for the yeast cytoplasmic threonyl-tRNA synthetase. *Nucleic Acids Res.*, **13**, 6171–83.

Pinsent, J. (1954). The need for selenite and molybdate in the formation of formic dehydrogenase by members of the *coli-aerogenes* group of bacteria. *Biochem. J.*, **57**, 10–16.

Powell, L. M., Wallis, S. C., Pease, R. J., Edwards, Y. H., Knott, T. J., and Scott, J. (1987). A novel form of tissue-specific RNA processing produces apolipoprotein-B48 in intestine. *Cell*, **50**, 831–40.

Prat, A., Katinka, M., Caron, F., and Meyer, E. (1986). Nucleotide sequence of the *Paramecium primaurelia* G surface protein. A huge protein with a highly periodic structure. *J. Mol. Biol.*, **189**, 47–60.

Preer, J. R., Jr., Preer, L. B., Rudman, B. M., and Barnett, A. J. (1985). Deviation from the universal code shown by the gene for surface protein 51A in *Paramecium. Nature*, **314**, 188–90.

Pritchard, A. E., Sable, C. L., Venuti, S. E., and Cummings, D. J. (1990*a*). Analysis of NADH dehydrogenase proteins, ATPase subunit 9, cytochrome b, and ribosomal protein L14 encoded in the mitochondrial DNA of *Paramecium. Nucleic Acids Res.*, **18**, 163–71.

Pritchard, A. E., Seilhamer, J. J., Mahalingam, R., Sable, C. L., Venuti, S. E., and Cummings, D. J. (1990*b*). Nucleotide sequence of the mitochondrial genome of *Paramecium. Nucleic Acids Res.*, **18**, 173–80.

Read, R., Bellew, T., Yang, J. G., Hill, K. E., Palmer, I. S., and Burk, R. F. (1990). Selenium and amino acid composition of selenoprotein P, the major selenoprotein in rat serum. *J. Biol. Chem.*, **265**, 17899–905.

Renbaum, P., Abrahamobe, D., Fainsod, A., Wilson, G. G., Rottem, S., and Razin, A. (1990). Cloning, characterization, and expression in *Escherichia coli* of the gene coding for the CpG DNA methylase from *Spiroplasma* sp. strain MQ1 (M. Sss1). *Nucleic Acids Res.*, **18**, 1145–52.

Roberts, R. B. (ed.) (1958). *Microsomal particles and protein synthesis*, (p. vi). The Lord Baltimore Press Inc., Washington DC.

Robertus, J. D., Ladner, J. E., Finch, J. T., Rhodes, D., Brown, R. S., Clark, B. F.C., and Klug, A. (1974). Structure of yeast phenylalanine tRNA at 3Å resolution. *Nature*, **250**, 546–51.

Roe, B. A., Wong, J. F. H., Chen, E. Y., and Armstrong, P. W. (1981). Sequence analysis of mammalian mitochondrial tRNAs. In *Recombinant DNA: Proceedings of the third Cleveland symposium on macromolecules* (ed. Walton, A. G.), pp. 167–76. Elsevier Science Publishing Co., Amsterdam.

Roe, B. A., Ma, D.-P., Wilson, R. K., and Wong, F.-H. (1985). The complete nucleotide sequence of the *Xenopus laevis* mitochondrial genome. *J. Biol. Chem.*, **260**, 9759–74.

Rould, M. A., Perona, J. J., Söll, D., and Steitz, T. A. (1989). Structure of *E. coli* glutaminyl-tRNA synthetase complexed with $tRNA^{Gln}$ and ATP at 2.8 Å resolution. *Science*, **246**, 1089–212.

Samuelsson, T., Guindy, Y. S., Lustig, F., Boren, T., and Lagerkvist, U. (1987). Apparent lack of discrimination in the reading of certain codons in *Mycoplasma mycoides. Proc. Natl. Acad. Sci. USA*, **84**, 3166–70.

Schimmel, P. (1987). Aminoacyl tRNA synthetases: General scheme of structure-function relationships in the polypeptides and recognition of transfer RNAs. *Ann. Rev. Biochem.*, **56**, 125–58.

Schimmel, P., Giegé, R., Moras, D., and Yokoyama, S. (1993). An operational RNA code for amino acids and possible relationship to genetic code. *Proc. Natl. Acad. Sci. USA*,

90, 8763–8.
Schimmel, P. R., Söll, D., and Abelson, J. N. (eds.) (1979). *Transfer RNA: structure, properties, and recognition*, (p. 519). Cold Spring Harbor Laboratory, New York.
Schneider, S. U., Leible, M. B., and Yang, X.-P. (1989). Strong homology between the small subunit of ribulose-1,5-biphosphate carboxylase/oxygenase of two species of *Acetabularia* and the occurrence of unusual codon usage. *Mol. Gen. Genet.*, **218**, 445–52.
Schulman, L. H. and Abelson, J. (1988). Recent excitement in understanding transfer RNA identity. *Science*, **240**, 1591–2.
Schultz, D. W. and Yarus, M. (1994). Transfer RNA mutation and the malleability of the genetic code. *J. Mol. Biol.*, **235**, 1377–80.
Schuster, W. and Brennicke, A. (1985). TGA-termination codon in the apocytochrome b gene from *Oenothera* mitochondria. *Curr. Genet.*, **9**, 157–63.
Schuster, W. and Brennicke, A. (1990). RNA editing of ATPase subunit 9 transcripts in *Oenothera* mitochondria. *FEBS Lett.*, **268**, 252–6.
Schuster, W., Hiesel, R., Wissinger, B., and Brennicke, A. (1990*a*). RNA editing in the cytochrome b locus of the higher plant *Oenothera berteriana* includes a U-to-C transition. *Mol. Cell Biol.*, **10**, 2428–31.
Schuster, W., Wissinger, B., Unseld, M., and Brennicke, A. (1990*b*). Transcripts of the NADH-dehydrogenase subunit 3 gene are differentially edited in *Oenothera* mitochondria. *EMBO J.*, **9**, 263–9.
Seilhamer, J. J. and Cummings, D. J. (1982). Altered genetic code in *Paramecium* mitochondria: possible evolutionary trends. *Mol. Gen. Genet.*, **187**, 236–9.
Shepherd, J. C. W. (1984). Fossil remnants of a primeval genetic code in all forms of life? *Trends Biochem. Sci.*, **9**, 8–10.
Shimayama, T., Himeno, H., Sasuga, J., Yokobori, S., Ueda, T., and Watanabe, K. (1990). The genetic code of a squid mitochondrial gene. *Nucleic Acids Res., Symp. Ser.*, **22**, 73–4.
Shimizu, M. (1982). Molecular basis of the genetic code. *J. Mol. Evol.*, **18**, 297–303.
Shinozaki, K., Ohme, M., Tanaka, M., Wakasugi, T., Hayashida, N., Matsubayashi, T., *et al.* (1986). The complete nucleotide sequence of the tobacco chloroplast genome: its gene organization and expression. *EMBO J.*, **5**, 2043–9.
Shuber, A. P., Orr, E. C., Recny, M. A., Schendel, P. F., May, H. D., Schauer, N. L., and Ferry, J. G. (1986). Cloning, expression, and nucleotide sequence of the formate dehydrogenase genes from *Methanobacterium formicicum*. *J. Biol. Chem.*, **261**, 12942–7.
Sibler, A. P., Dirheimer, G., and Martin, R. P. (1981). Nucleotide sequence of a yeast mitochondrial threonine-tRNA able to decode the CUN leucine codons. *FEBS Lett.*, **132**, 344–8.
Sibler, A. P., Dirheimer, G., and Martin, R. P. (1985). Yeast mitochondrial $tRNA^{Ile}$ and $tRNA^{Met}_{m}$: nucleotide sequence and codon recognition patterns. *Nucleic Acids Res.*, **13**, 1341–5.
Sibler, A. P., Dirheimer, G., and Martin, R. P. (1986). Codon reading patterns in *Saccharomyces cerevisiae* mitochondria based on sequences of mitochondrial tRNAs. *FEBS Lett.*, **194**, 131–8.
Simoneau, P. , Li, C.-M., Loechel, S., Wenzel, R., Herrmann, R., and Hu, P.-C. (1993). Codon reading scheme in *Mycoplasma pneumoniae* revealed by the analysis of the complete set of tRNA genes. *Nucleic Acids Res.*, **21**, 4967–74.
Smith, M. J., Banfield, D. K., Doteval, K., Gorshi, S., and Kowbel, D. J. (1990). Nucleotide sequence of nine protein-coding genes and 22 tRNAs in the mitochondrial DNA of the sea star *Pisaster ochraceus*. *J. Mol. Evol.*, **31**, 195–204.
Söll, D. (1988). Enter a new amino acid. *Nature*, **331**, 662–3.

Söll, D., Cherayil, J., Jones, D. S., Faulkner, R. D., Hampel, A., Bock, R. M., and Khorana, H. G. (1966*a*). sRNA specificity for codon recognition as studied by the ribosomal binding technique. *Cold Spring Harbor Symp. Quant. Biol.*, **31**, 51–61.

Söll, D., Jones, D. S., Ohtsuka, E., Faulkner, R. D., Lohrmann, R., Hayatsu, H., and Khorana, H. G. (1966*b*). Specificity of sRNA for recognition of codons as studied by the ribosomal binding technique. *J. Mol. Biol.*, **19**, 556–73.

Sommer, B., Kohler, M., Sprengel, R., and Seeburg, P. H. (1991). RNA editing in brain controls a determinant of ion flow in glutamate-gated channels. *Cell*, **67**, 11–19.

Speyer, J., Lengyel, P., Basilio, C., Wahba, A., Gardner, R., and Ochoa, S. (1963). Synthetic polynucleotides and the amino acid code. *Cold Spring Harbor Symp. Quant. Biol.*, **28**, 559–67.

Sprinzl, M., Hartmann, T., Weber, J., Blank, J., and Zeidler, R. (1989). Compilation of tRNA sequences and sequences of tRNA genes. *Nucleic Acids Res.*, **17**(Suppl.), 1–173.

Stadtman, T. C. (1990). Selenium biochemistry. *Ann. Rev. Biochem.*, **59**, 111–27.

Steitz, J. A. (1968). Nucleotide sequence of the ribosomal binding sites of bacteriophage R17 RNA. *Cold Spring Harbor Symp. Quant. Biol.*, **34**, 621–30.

Stern, D. B., Bang, A. G., and Thompson, W. F. (1986). The watermelon mitochondrial URF-1 gene: Evidence for a complex structure. *Curr. Genet.*, **10**, 857–69.

Stretton, A. O. W., Kaplan, S., and Brenner, S. (1966). Nonsense codons. *Cold Spring Harbor Symp. Quant. Biol.*, **31**, 173–9.

Sturchler, C., Westhof, E., Carbon, P., and Krol, A. (1993). Unique secondary and tertiary structural features of the eucaryotic selenocysteine $tRNA^{Sec}$. *Nucleic Acids Res.*, **21**, 1073–9.

Sueoka, N. (1961). Correlation between base composition of deoxyribonucleic acid and amino acid composition of protein. *Proc. Natl. Acad. Sci. USA*, **47**, 1141–9.

Sueoka, N. (1962). On the genetic basis of variation and heterogeneity of DNA base composition. *Proc. Natl. Acad. Sci. USA*, **48**, 582–92.

Sueoka, N. (1988). Directional mutation pressure and neutral molecular evolution. *Proc. Natl. Acad. Sci. USA*, **85**, 2653–7.

Sueoka, N. (1993). Directional mutation pressure, mutator mutations, and dynamics of molecular evolution. *J. Mol. Evol.*, **37**, 137–53.

Suzuki, T., Ueda, T., Yokogawa, T., Nishikawa, K., and Watanabe, K. (1994). Characterization of serine and leucine tRNAs in an asporogenic yeast *Candida cylindracea* and evolutionary implications of genes for $tRNA^{Ser}_{CAG}$ responsible for translation of a non-universal genetic code. *Nucleic Acids Res.*, **22**, 115–23.

Szathmáry, E. (1993). Coding coenzyme handles: A hypothesis for the origin of the genetic code. *Proc. Natl. Acad. Sci. USA*, **90**, 9916–20.

Takanami, M. and Yan, Y. (1965). The release of polypeptide chains from ribosomes in cell-free amino acid-incorporating systems by specific combinations of bases in synthetic polyribonucleotides. *Proc. Natl. Acad. Sci. USA*, **54**, 1450–5.

Tanaka, R., Muto, A., and Osawa, S. (1989). Nucleotide sequence of tryptophan tRNA gene in *Acholeplasma laidlawii. Nucleic Acids Res.*, **17**, 5842.

Terzaghi, E., Okada, Y, Streisinger, G., Emrich, J., Inoue, M., and Tsugita, A. (1966). Change of a sequence of amino acids in phage T4 lysozyme by acridine-induced mutations. *Proc. Natl. Acad. Sci. USA*, **56**, 500–7.

Tham, T. N., Ferris, S., Kovacic, R., Montagnier, L., and Blanchard, A. (1993). Identification of *Mycoplasma pirum* genes involved in the salvage pathways for nucleosides. *J. Bacteriol.*, **175**, 5281–5.

Tormay, P., Wilting, R., Heider, J., and Böck, A. (1994). Genes coding for the selenocysteine-inserting tRNA species from *Desulfomicrobium baculatum* and *Clostridium thermoaceticum*: Structural and evolutionary implications. *J. Bacteriol.*, **176**, 1268–74.

Ueda, T. and Watanabe, K. (1993). The evolutionary change of the genetic code as restricted by the anticodon and identity of transfer RNA. *Origins of Life and Evol. Biosph.*, **23**, 345–64.

Umesono, K. and Ozeki, H. (1987). Chloroplast gene organization in plants. *Trends Genet.*, **3**, 281–7.

Vold, B. (1985). Structure and organization of genes for transfer ribonucleic acid in *Bacillus subtilis. Microbiol. Rev.*, **49**, 71–80.

Volkin, E. and Astrachan, L. (1956). Phosphorus incorporation in *Escherichia coli* ribonucleic acid after infection with bacteriophage T2[1]. *Virology*, **2**, 149–61.

Wagner, R. B. and Zoo, H. D. (1953). *Synthetic organic chemistry*, pp. 479–532. Wiley, New York.

Wakasugi, T., Ohme, M., Shinozaki, K., and Sugiura, M. (1986). Structure of tobacco chloroplast genes of tRNA Ile(CAU), tRNA Leu(CAA), tRNA Cys(GCA), tRNA Ser(UGA) and tRNA Thr(GGU): a compilation of tRNA genes from tobacco chloroplasts. *Plant Mol. Biol.*, **7**, 385–92.

Waring, R. B., Davies, R. W., Lee, S., Grisi, E., Berks, M. M., and Scazzocchino, C. (1981). The mosaic organization of the apocytochrome b gene of *Aspergillus nidulans* revealed by DNA sequencing. *Cell*, **27**, 4–11.

Warrick, H. and Spudich, J. A. (1988). Codon preference in *Dictyostelium discoideum. Nucleic Acids Res.*, **14**, 6617–35.

Watanabe, K., Ueda, T., Yokogawa, T., Suzuki, T., Nishikawa, K., Mori, M., *et al.* (1993). Molecular mechanism of the genetic code variations found in *Candida* species and its implications in evolution of the genetic code. In *The translational apparatus* (ed. Nierhaus *et al.*) pp. 647–56. Plenum Press, New York.

Weber, A. L. and Miller, S. L. (1981). Reasons for the occurrence of the twenty coded protein amino acids. *J. Mol. Evol.*, **17**, 273–84.

Weber, F., Dietrich, A., Weil, J.-H., and Marechal-Drouard, L. (1990). A potato mitochondrial isoleucine tRNA is coded for by a mitochondrial gene possessing a methionine anticodon. *Nucleic Acids Res.*, **18**, 5027–30.

Webster, R. E., Engelhardt, D.L., and Zinder, N. D. (1966). *In vitro* protein synthesis: chain initiation. *Proc. Natl. Acad. Sci. USA*, **55**, 155–61.

Weiner, A. M. (1987). Evolution of the gene. In *Molecular biology of the gene*, 4th ed., Vol. 2, (ed. Watson, J. D. *et al.*), pp. 1097–163, Benjamin/Cummings Publishing Company, Inc., California.

Weiss, W. A. and Friedberg, E. C. (1986). Normal yeast $tRNA^{Gln}$(CAG) can suppress amber codons and is encoded by an essential gene. *J. Mol. Biol.*, **192**, 725–35.

Wilcox, M. (1969). γ-phosphoryl ester of Glu-$tRNA^{Gln}$ as an intermediate in *Bacillus subtilis* glutaminyl-tRNA synthesis. *Cold Spring Harbor Symp. Quant. Biol.*, **34**, 521–33.

Williams, K. R. and Herrick, G. (1991). Expression of the gene encoded by a family of macromolecular chromosomes generated by alternative DNA processing in *Oxytricha fallax. Nucleic Acids Res.*, **19**, 4717–24.

Wintz, H., Grienenberger, J.-M., Weil, J.-H., and Lonsdale, D. M. (1988). Location and nucleotide sequence of two tRNA genes and a tRNA pseudo-gene in the maize mitochondrial genome: evidence for the transcription of a chloroplast gene in mitochondria. *Curr. Genet.*, **13**, 247–54.

Wissinger, B., Schuster, W., and Brennicke, A. (1991). Trans-splicing in *Oenothera* mitochondria: nad 1 mRNAs are edited in exon and trans-splicing group II intron sequences. *Cell*, **65**, 473–82.

Woese, C. R. (1969). Models for the evolution of codon assignments. *J. Mol. Biol.*, **34**, 235–40.

Wolff, G., Plante, I., Lang, B. F., Kück, U., and Burger, G. (1994). Complete sequence of the mitochondrial DNA of the chlorophyte alga *Prototheca wickerhamii*. Gene content and genome organization. *J. Mol. Biol.*, **237**, 75–86.

Wolstenholme, D. R. (1992). Animal mitochondrial DNA: structure and evolution. In *Mitochondrial genomes* (ed. Jeon, K. W. and Wolstenholme, D. R.). *Intern. Rev. Cytol.*, **141**, 173–216.

Wolstenholme, D. R., Macfarlane, J. L., Okimoto, R., Clary, D. O., and Wahleithner, J. A. (1987). Bizarre tRNAs inferred from DNA sequences of mitochondrial genomes of nematode worms. *Proc. Natl. Acad. Sci. USA*, **84**, 1324–8.

Wolstenholme, D. R., Okimoto, R., Macfarlane, J. L., Pont, G. A., Chamberlin, H. M., Garey, J. R., and Okada, N. A. (1990). Unusual features of lower invertebrate mitochondrial genomes. In *Structure, function and biogenesis of energy transfer systems* (ed. Quagriello, E., Papa, S., Palmieri, F., and Saccone, C.), pp. 103–6, Elsevier Biomedical Press, Amsterdam.

Wong, J. T.-F. (1975). A co-evolution theory of the genetic code. *Proc. Natl. Acad. Sci. USA*, **72**, 1909–12.

Wong, J. T.-F. (1976). The evolution of a universal genetic code. *Proc. Natl. Acad. Sci. USA*, **73**, 2336–40.

Wong, J. T.-F. (1981). Coevolution of genetic code and amino acid biosynthesis. *Trends Biochem. Sci.*, pp. 33–5.

Wong, J. T.-F. (1988). Evolution of the genetic code. *Microbiol. Sci.*, **5**, 174–81.

Wong, J. T.-F. and Cedergren, R. (1986). Natural selection versus primitive gene structure as determinant of codon usage. *Eur. J. Biochem.*, **159**, 175–80.

Yamao, F., Muto, A., Kawauchi, Y., Iwami, M., Iwagami, S., Azumi, Y., and Osawa, S. (1985). UGA is read as tryptophan in *Mycoplasma capricolum*. *Proc. Natl. Acad. Sci. USA*, **82**, 2306–9.

Yamao, F., Iwagami, S., Azumi, Y., Muto, A., Osawa, S., Fujita, N., and Ishihama, A. (1988). Evolutionary dynamics of tryptophan tRNAs in *Mycoplasma capricolum*. *Mol. Gen. Genet.*, **212**, 364–9.

Yamao, F., Andachi, Y., Muto, A., Ikemura, T., and Osawa, S. (1991). Levels of tRNAs in bacterial cells as affected by amino acid usage in proteins. *Nucleic Acids Res.*, **19**, 6119–22.

Yang, A. J. and Mulligan, R. M. (1989). RNA editing intermediates of cox 2 transcripts in maize mitochondria. *Mol. Cell. Biol.*, **11**, 4278–81.

Yarus, M. (1988). tRNA identity: A hair of the dogma that bit us. *Cell*, **55**, 739–41.

Yarus, M. (1991). An RNA-amino acid complex and the origin of the genetic code. *New Biologist*, **3**, 183–9.

Yokobori, S., Ueda, T., and Watanabe, K. (1993). Codons AGA and AGG are read as glycine in ascidian mitochondria. *J. Mol. Evol.*, **36**, 1–8.

Yokogawa, T., Suzuki, T., Ueda, T., Mori, M., Ohama, T., Kuchino, Y., *et al.* (1992). Serine tRNA complementary to the nonuniversal serine codon CUG in *Candida cylindracea*: evolutionary implications. *Proc. Natl. Acad. Sci. USA*, **89**, 7408–11.

Yokoyama, S., Watanabe, T., Murao, K., Ishikura, H., Yamaizumi, Z., Nishimura, S., and Miyazawa, T. (1985). Molecular mechanism of codon recognition by tRNA species with modified uridine in the first position of the anticodon. *Proc. Natl. Acad. Sci. USA*, **82**, 4905–9.

Yoshida, M., Takeishi, K., and Ukita, T. (1970). Anticodon structure of GAA specific glutamic acid tRNA from yeast. *Biochem. Biophys. Res. Commun.*, **39**, 852–7.

Zamaroczy, M. and Bernardi, G. (1986). The GC clusters of the mitochondrial genome and their evolutionary origin. *Gene*, **41**, 1–22.

Zamecnik, P. C. (1969). An historical account of protein synthesis, with current overtones – A personalized view. *Cold Spring Harbor Symp. Quant. Biol.*, **34**, 1–16.

Zinoni, F., Birkmann, A., Stadtman, T. C., and Böck, A. (1986). Nucleotide sequence and expression of selenocysteine-containing polypeptide of formate dehydrogenase (formate-hydrogen-lyase linked) from *Escherichia coli*. *Proc. Natl. Acad. Sci. USA*, **83**, 4650–4.

Zinoni, F., Heider, J., and Böck, A. (1990). Features of the formate-dehydrogenase mRNA necessary for decoding of the UGA codon as selenocysteine. *Proc. Natl. Acad. Sci. USA*, **87**, 4660–4.

Index

Abbreviations: The RNA bases are indexed under their abbreviated form e.g. A, C, G, U, whether alone or as part of a sequence (codon/anticodon). Amino acids are entered under their full form when the first word of main headings, but otherwise are used in their standard abbreviated forms.

A 22, 27
  base-pairing rules 16, 22
  end anticodon position 26
  first anticodon position 22
  first codon position 22, 27
  at nucleotide 37 in tRNA, modified 27
  RNA editing to G 132
AAA codon 88–9, 94, 113–15, 174–5, 175–6
  Asn-coding 85, 88, 93, 113–15, 174–5, 175–6
  identification 7
AAG codon 89
abiotic synthesis of amino acid 133–4
ACN codon, yeast mitochondrial 106
adaptor hypothesis 5–6
adenosine, *see* A *and codons/anticodons containing A*
AGA codon 174
  Ser-coding 89
  stop coded by 91, 111
  unassigned 65, 111
AGG codon 174
  stop coded by 91, 111
AGR codon 89–91, 93, 108–13
  Ser/Gly-coding 85, 89–91, 93, 108–13
  stop coded by 85, 89–91, 93, 108–13
AGU anticodon 22, 23
alanine (Ala)
  protein content 163
  universal codons for 10
    usage 60
amino acid(s)
  abiotic synthesis 133–4
  attachment to tRNA 31–5
  codons for 14
    deciphering 7–8
    synonymous, *see* synonymous codons
    unassigned codons arising from, *see* unassigned codons
    universal 14
    see also specific codons
  composition in protein 161–70
  phase 1 or 2, code evolution and 149, 150
  21st, selenocysteine as 124
  see also specific amino acids
aminoacylation (of tRNA)
  evolution of code and 139–41
  suppressor anticodon insertion and its effects on 136, 137
aminoacyl-tRNA synthetases (ARS) 31–5, 147, 148
  evolution of code and 139–45, 146, 147, 148, 149, 151
  mitochondrial 34–5
  phylogenetic trees 147, 148, 149, 150
  see also specific synthetases
anticodons 36–57
  of archetypal/early code 151–2, 154
  base-pairing rules with codons, *see* base-pairing rules
  composition 36–57, 62, 64–5
  deletion, *see* deletion
  evolutionary diversification 157–8
  suppressor, aminoacylation and effects of insertion of 136, 137
  see also tRNA; *the specific anticodon species; specific anticodons; specific amino acids*
archaebacteria
  anticodons in 41
  nucleoside 37 of tRNA in 30
arginine (Arg)
  ICG anticodon for 158
  protein content 162–3
  universal codons for 10, 59
    Ser/Gly-coding 85
    stop coded by 85
    Trp-coding 126–7
    usage 60, 61
A-site of ribosome 69
asparagine (Asn) codons 10, 88–9, 174–5, 175–6
  AAA as 88–9, 113–15, 174–5, 175–6

asparagyl-tRNA synthetase 34
aspartic acid (Asp), universal codons for 10
Aspergillus mitochondria, non-universal code 104, 105, 106, 107
AT codons (AA/AT/TA/TT in first two positions), amino acids assigned by 165, 166
AT pressure, codon non-use and 59–61
AUA codon 88, 113–15, 157
  conversion to AUG 156
  deletion and reassignment 96
  disappearance 96
  Ile-coding, evolution 157
  Met-coding 24, 25, 84, 85, 88, 106–7, 113–15
    in early code 153, 156
    in early to universal code evolution 156
    reversal to Ile 113–15, 157
  unassigned 65
AUG codon
  initiation coded by 9, 50
  Met-coding 24, 25
    in early code 153
    in early to universal code evolution 156
AUN codon changes during evolution 95–6

Bacillus subtilis, nucleoside 37 of tRNA in 28–9
bacteria
  anticodon composition 36–7
  codon usage and G + C content 46–51
  see also specific genera/species
base pairing rules (codon–anticodon) 11, 15–25, 62–3, 64–5
  current status 16
  nucleotide 37 effects 26–31

C (cytidine) 24–5, 27
  base-pairing rules 16, 24–5
  end anticodon position 31
  first anticodon position 24–5
  first codon position 24–5, 27
  in RNA editing
    insertion 132
    U substituted by 126
    U substituting for 126, 127–9, 131, 132
  see also G + C content; GC codons; GC pressure
C (deoxycytidine)
  reversion to T, RNA editing and 130, 130–1
  T mutating to, RNA editing and 129
*C 24
Candida spp., CUG codon as Ser 72, 79, 80
capture, codon 93–6, 97–9, 171, 173, 174
5-carboxymethylaminomethyl-2′-O-methyluridine ($cmnm^5Um$) 19
5-carboxymethylaminomethyl-2-thiouridine ($cmnm^5s^2U$) 19
5-carboxymethylaminomethyluridine ($cmnm^5U$) 20
CCC codon, identification 7
CGG codon 59, 63–4, 126–7, 174
  translation ceasing just before 63–4, 68
  Trp-coding 126–7
  unassigned 59, 66
CGN codon 109, 110
chloroplasts
  anticodons in 37, 42
  evolution of code 158
  RNA editing 131
ciliated protozoans, non-universal codons 77–9
$cmnm^5s^2U$ 19
$cmnm^5U$ 20
  base-pairing rules 16, 20
$cmnm^5Um$ 19
cmo^5U 19
CNN anticodon 21, 36–7, 157
coding, fundamental features 1
coding co-enzyme hypothesis 145–7
codons 1, 14–31, 45–115
  ambiguous assignment (simultaneously with more than one meaning) 123, 175–6
  amino acid, *see* amino acids
  base-pairing with anticodons, *see* base-pairing rules
  capture 93–6, 97–9, 171, 173, 174
  deletion/disappearance, *see* deletion
  of early code 153–5
  identification 7–11
  initiation, *see* initiation codons
  non-universal 74–115
    mechanism of change/reassignment 93–6
  selection 53–6
    isoacceptor tRNA amount related to 58, 60
    negative 53, 54, 68
    positive 53–4, 55–6
  synonymous, *see* synonymous codons
  termination, *see* termination codons
  triplet nature 3, 6–7
  unassigned/nonsense, *see* unassigned codons
  usage 45–115
  see also specific codons
commaless code 3–5, 7, 137
CUG codon, Ser-coding 72, 79–80, 81, 99–103
  reversal to Leu 99–103
CUN codon, Thr-coding 85, 91–2, 103–6
CUR codons, fungal mitochondrial 104
cysteine (Cys) codons
  UGA as 72, 77–8, 82, 99
  universal 10
    selenocysteine-coding UGA changing to 124
cytidine, *see C and codons/anticodons containing C*

cytochrome oxidase genes
  AAA/AAG codons 89
  AGA/AGG codons 91
  RNA editing sites 127, 130
  UGG codon 88

deletion/disappearance
  of codons/anticodons 58–62
    temporary 93, 96, 115
  of U or C, RNA editing involving 132
diamond code 3
directional mutation 48, 165–8
  asymmetrical 52
  pressures 52, 61–3, 165–8
DNA, historical perspectives 2
doublet code 3

early evolution of genetic code 133–60
editing, RNA, *see* RNA
elongation factor of selenocysteyl-tRNA$^{(Ser)Sec}$ (SELB) 119, 121
Escherichia coli
  amino acid composition in 166–8, 169, 170
    tRNA content and 169
  anticodons in 28–9
  codon usage 49, 50
  mut genes 45
  nucleoside 37 of tRNA in 28–9
eubacteria
  anticodons 38–9, 62–3, 64–5
  codon–anticodon pairing patterns 62–3, 64–5
  nucleoside 37 of tRNA in 28–9, 30
  Trp codon usage 77
eukaryotes
  anticodons in 40, 44
  codon selection/usage 51–2
    G + C content and 51–2
    stop 57
  nucleoside 37 of tRNA in 30
  release factor (eRF) 57
  RNA editing in nucleus of 132

family (four-codon box) box 14
  code evolution and 151
  synonymous codon selection in 56
f^5C 88
FCo frame-shift mutant 6
fixed reading frames 6
5-formylcytidine (f^5C) 88
four-codon box, *see* family box
four-way wobbling 25–6
frame-shift mutant studies 6
free site (in synonymous codons) 14
frozen-accident theory 71
fungi
  mitochondria, anticodons in 38–9, 43
  non-universal code 103–6
  see also yeast

G 17, 31
  base-pairing rules 16, 17
  end anticodon position 27
  first anticodon position 17
  first codon position 17, 31
  RNA editing from A to 132
GAU anticodon converted to IAU 160
G + C content
  amino acid composition of proteins and 164, 165, 166
  anticodon composition and 62, 64–5
  codon usage and 45–52, 55–6
GC codons (GG/GC/CG/CC in first two positions), amino acids assigned by 165, 166
GC pressure
  amino acid composition of proteins and 165, 166, 168
  T-to-C mutation and C-to-T reversion related to 130
GCU anticodon 111
GGG codon, identification 7
glutamic acid (Glu) codons 10
  usage 60
glutamine (Gln) codons
  UAR as 72, 77, 78, 99
  universal 10
  usage 60
glutaminyl-tRNA synthetase 32
  code evolution and 151
glycine (Gly) codons 10, 89–91, 108–13
  AGR as 89–91, 108–13
  GGG as, identification 7
  universal 10
  usage 60
GNN anticodon 36–7, 157
guanosine, *see* G *and codons/anticodons containing G*
GUG codon
  initiation coded by 9, 50
  Val-coding 9

H strands of mitochondrial genome, G + C content 52
histidine (His) universal codons 10
5-hydroxyuridine derivatives (xo^5U) 16, 18–19

I (inosine) 17, 21–2, 158–60
  base-pairing rules 16
  evolution of anticodons with 158–60
initiation codons 9
  bacterial 50
  G + C content and 50
  see also specific codons
INN anticodon evolution 158–60
inosine, *see* I
insertion/deletion editing 132

isoleucine (Ile)
  Met-coding AUA codon reverting to 113
  universal codons for 10, 157
    evolution 157
    Met-coding 85, 88, 106–7, 113–15, 156
    usage 60, 61
isoleucyl-tRNA synthetase (IleRS) 32, 33–4

L (2-lysyl C; lysidine) 23, 24
  base-pairing rules 16, 23
L-shaped tRNA tertiary structure 13, 86
L strands of mitochondrial genome, G+C content 52
leucine (Leu)
  Ser CUG codon reverting to 99–103
  universal codons for 10
    Ser-coding 72, 79–80, 81, 99–103
    Thr-coding 85, 91–2, 103–6
    usage 60, 61
lysine (Lys)
  protein content 162–3
  universal codons for
    AAA, *see* AAA
    Asn-coding 85, 88–9, 113–15, 174–5, 175–6
    identification 7
    usage 60, 61
2-lysyl C, *see* L

mcm^5s^2U 19
mcm^5U 20
  base-pairing rules 16, 20
messenger RNA, *see* RNA
metabacteria, nucleoside 37 of tRNA in 30
meteorites, amino acids on 150
methionine (Met) codons 10, 88, 156
  AUA/AUG, *see* AUA; AUG
  reduction in numbers 156
  universal 10
5-methoxycarbonylmethyl-2-thiouridine (mcm^5s^2U) 19
5-methoxycarbonylmethyluridine (mcm^5U) 20
5-methoxyuridine (mo^5U) 19
5-methylaminomethyl-2-thiouridine (mnm^5s^2U) 19
5-methylaminomethyluridine (mnm^5U) 19
2′-O-methyl pyrimidine derivatives 20
5-methyl-2-thiouridine derivatives (xm^5s^2U) 19
2′-O-methyluridine (Um) 16, 20
Micrococcus luteus
  amino acid composition in 166–8, 169, 170
    tRNA content and 169
  anticodons in 38–9
  codons
    unassigned 65, 66
    usage/selection 49, 50, 51, 56, 58, 60
  nucleoside 37 in tRNA of 28–9
mitochondria 34–5, 37–40, 52, 103–15, 126–31, 132
  aminoacyl-tRNA synthetases 34–5
  anticodons in 37–40, 43
    code evolution and 153–5
  codons 82–93, 103–15
    non-universal 73, 82–93, 103–15
    unassigned 63, 73
    usage 52, 59
  evolution of code and phylogenetic tree 86, 92–3, 158
  Mycoplasma spp. and, resemblance between 37–40
  RNA editing 126–31, 132
  tRNA genes in *S. cerevisiae*, base composition 59, 62
  $tRNA^{Arg}$ 26
  $tRNA^{Trp}$ 87, 88
  vertebrate, universal genetic code differing from code in 71
mnm^5s^2U 19
mnm^5U 19
mo^5U 19
mRNA (messenger RNA)
  discovery 7
  evolution 141, 145
mut genes 45
mutation
  directional, *see* directional mutation
  pressures 61–3, 165–8
    G+C content and 47–8, 52, 165–8
  replacement site 14–15
    conservative 15
  RNA editing and 129, 130
Mycoplasma spp.
  A (first anticodon nucleoside) base-pairing 22
  anticodon composition 36–7, 38–9
  codons in
    non-universal 74–5, 96–7
    unassigned 58–61 *passim*, 66
    usage 49, 50, 58–9, 61
  G (first anticodon nucleoside) base-pairing 17
  mitochondria and, resemblance between 37–40
  tRNA nucleoside 37 in 28–9
  $tRNA^{Thr}$ 26
  $tRNA^{Trp}$ 96–7
  U (first anticodon nucleoside) base-pairing 17

NAR codon 17
NAY codon 17
Neurospora mitochondria, non-universal code 104, 105, 106, 107
NNA codon 21, 58–9
NNC codon 58, 59
NNG codon 21, 58–9
  selection by tRNA 54
NNR codon 19, 58
  selection by tRNA 54

NNU codon 58–9
  selection by tRNA 19
NNY codon
  in early code 155
  selection by tRNA 19
nonsense codons, *see* unassigned codons
non-universal code 71–125, 171–6
  mechanisms of change/reassignment 93–6
nuclear systems
  non-universal code 72, 74–82, 96–103
  RNA editing 132
nucleotide 37 in tRNA, codon–anticodon pairing and 26–31

origin of genetic code 133–53
overlapping code 2

P-site of ribosome 69
peptidyl transferase reaction 145
peptidyl-tRNA 69–70
phenylalanine (Phe) universal codons 10
  UUU, identification 7
phylogenetic trees/relationships
  aminoacyl-tRNA synthetases 147, 148, 149, 150
  in code evolution 172
  mitochondria 86, 92–3
  yeast 79, 84
plant 126–31
  chloroplast, anticodons in 37, 42
  mitochondria 126–31
    L-shaped tRNA tertiary structure 86
    RNA editing 126–31
    universal code 82
proline (Pro) universal codons 10
  CCC, identification 7
  usage 60
protein
  amino acid composition 161–70
  synthesis (translation)
    blockage by unassigned codons 63–4, 68
    evolution 137–9, 141–4, 151, 159
Prototheca wickerhamii mitochondrial unassigned codons 59, 66

Q (queuosine), base-pairing rules 16, 17

reading frame, fixed 6
release factors
  eukaryotic RF (eRF) 57
  stop codon selection by 56
  type 1 (RF-1) 56, 57
  type 2 (RF-2) 25, 56–7
    gene deletion in *M. capricolum* 97
replacement site 14–15
RF, *see* release factors
ribosomal RNA, ribosome function and 144–5
ribosome, evolution 141, 144
ribozyme origin of code 141–7
rII region of T4, early studies of 6
RNA
  editing 126–32, 176–7
    biological meaning 126
    model for evolution 127–31
  historical perspectives 2
  operational code, evolution of genetic code from 139–41
RNY hypothesis of code origin 137–9
rRNA (ribosomal RNA), ribosome function and 144–5
RRY hypothesis of code origin 137–9

Saccharomyces cerevisiae mitochondria
  non-universal code 103, 104, 105
  tRNA genes, base composition 59, 62
  unassigned codons 59
Sel A 118
Sel B 119
Sel C 118
Sel D 119
selection 162–8
  against genetic code 162–8
  codon, *see* codon
selenocysteine 116–32, 176
  biosynthesis 118–19, 124–5
  UGA coded by 116–32, 176
selenocysteyl-$\text{tRNA}^{(\text{Ser})\text{Sec}}$ 118, 119, 120, 121
selenoprotein-P 116, 123
Sel-P 116, 123
serine (Ser) codons 10, 89–91, 99–103, 108–13
  AGR as 89–91, 108–13
  CUG as, *see* CUG
  universal 10
start codons, *see* initiation codons
stereochemical mechanisms of code origin 135, 136–7, 149
stop codons, *see* termination codons
str operon, amino acid composition coded by 166–8
suppressor anticodon insertion, aminoacylation effects of 136, 137
suppressor strains, stop codons and 8
synonymous codons 14, 53
  usage/selection of 55–6
    corresponding tRNA/anticodon and, relationship between 58

T
  C reverting to, RNA editing and 130, 130–1
  mutation to C, RNA editing and 129, 130
  see also AT codons; AT pressure
T4 rII region, early studies of 6

template hypothesis 1–2
termination (stop) codons 8–9, 10, 25, 63–70, 96–9
  amino acid-coding 72, 77, 86–8, 96–9
    Cys 72, 82, 99
    Gln 72, 77, 99
    reversal to stop 156
    selenocysteine 116–32
    Trp 72, 84, 85, 86–8, 92–3, 96–9
    Tyr 85, 92, 113–15
  amino acid universal codons as 85, 89–91, 93, 108–13
  bacteria 50–1
  capture 97–9
  G+C content and 50–1
  nucleotide following, importance for efficient termination 57
  selection by release factors 56
  selenocysteine codon changing to 124
  unassigned codons and 63–70
  see also specific codons
2-thiouridine derivatives 20; *see also specific derivatives*
threonine (Thr) codons
  CUN 91–2, 103–6
  universal 10
    usage 60
thymidine, *see* AT pressure; T
Torulopsis glabrata mitochondria
  non-universal code 105
  unassigned codons 59, 66
transfer RNA, *see* RNA
translation, *see* protein
triplet nature of codon 3, 6–7
tRNA (transfer/soluble RNA) 5–6, 11–13, 31–5, 53–6, 147, 168–70
  as adaptor 5–6
  amino acid attachment 31–5
  amino acid usage in proteins affecting levels of 168–70
  anticodon, *see* anticodon
  codon selection by 53–6
    negative 53, 54
    positive 53–4, 55–6
  evolution 141–4, 147
  historical perspectives 5–6
  isoacceptor, codon usage related to amount of 58, 60
  mitochondrial *S. cerevisiae*, base composition of genes 59, 62
  nucleotide 37 of, codon–anticodon pairing and 26–31
  structure 12–13
$tRNA^{Arg}$ 22
  with anticodon UCU 109, 110
  yeast mitochondrial 26
$tRNA^{Asn}$ 88–9, 175
  amino acid charging 34
$tRNA^{Gln}$ 21, 32, 78–9
$tRNA^{Gly}$ 17, 22, 108, 111
  with anticodon UCC 111
  with anticodon UCU 108, 111
$tRNA^{Ile}$, amino acid charging 32, 33–4
$tRNA^{Leu}$, UAG species 103, 104, 106
$tRNA^{Lys}$ 115, 175, 176
$tRNA^{Met}$ 88
  with anticodon CAU 9, 88, 107, 115
$tRNA^{Met\text{-}i}$ (initiator) 27
$tRNA^{Met\text{-}m}$ (elongator) 27
$tRNA^{Ser}$ 22, 89–91, 101
  with anticodon CAG 40, 79, 82, 83, 101, 102
  with anticodon CGA 102
  with anticodon GCU 89–91, 108, 110
  with anticodon IGA 79, 83, 102
  $tRNA^{(Ser)Sec}$ structure compared with 118, 119, 120
$tRNA^{(Ser)Sec}$ 116, 117, 118, 124, 125
  gene 116, 117, 118
  structure 118, 119–20
$tRNA^{Thr}$ 22, 91–2, 96–7, 110
  with anticodon AGU 26
  with anticodon CGU 17
  with anticodon UAG 17, 92, 104, 106
  with anticodon UGU 17, 92
$tRNA^{Trp}$ 96–7, 98, 110
  with anticodon CCA 76, 96–7
  genes 96, 98
  mitochondrial 87, 88
  with anticodon UCA 76, 96, 99
tryptophan (Trp) codons 10, 86–8, 96–9, 103, 156
  CGG as 126–7
  reduction in numbers 156
  UGA/UGG as, *see* UGA; UGG
  universal 103
    usage 61
two-codon set 14
  code evolution and 151
  codon selection 55–6
two-out-of-three hypothesis 25–6, 70
tyrosine (Tyr) codons
  UAA as 92, 113–15
  universal 10

U 17–21, 26–7
  base-pairing rules 16, 17–21, 26–7
  end anticodon position 27
  first anticodon position 18–20
  first codon position 26–7
  modified derivatives 19
  in RNA editing
    C substituted by 126, 127–9, 131, 132
    C substituting for 126
    insertion/deletion 132
+U 18
  base-pairing rules 18–19
*U 20
  base-pairing rules 20–1

UAA codon 113–15, 174
  stop coded by 8, 25, 50, 51
    selection by release factor 56–7
  Tyr-coding 85, 92, 113–15
UAG codon 174
  conversion to AGR stop codon 112–14
  stop coded by 8, 25, 50, 51
UAR codon, Gln-coding 72, 77, 78, 99
*UCA anticodon 116, 123, 156
*UCU anticodon 21
UGA codon 74–7, 86–8, 96–9, 116–32, 156
  conversion to UGG 156
  Cys-coding 72, 77–8, 99
  in early to universal code evolution 156
  selenocysteine-coding 116–32, 176
  stop coded by 8, 25, 50, 51, 156
    selection by release factor 56–7
  Trp-coding 72, 74–7, 84, 85, 86–8, 92–3, 96–9, 103
    in early code 153
    in early to universal code evolution 156
    reversal to stop 156
  unassigned codon 59–60, 156
UGG codon as Trp 88
  in early code 153
  in early to universal code evolution 156
UGY, selenocysteine UGA codon changing to 124
Um, base-pairing rules 16, 20
unassigned (nonsense) codons 58–70, 73, 171–3
  capture 93–6, 97–9, 171, 173, 174
  genesis 58–63
  reassignment 93–6, 97–9, 173–6
  stop codons and 63–70
universal genetic code 14–35, 155–60
  evolution 155–60
  vertebrate mitochondrial code differing from 71
UNN anticodon in early code 155
*UNN anticodon 20, 21, 59
  deletion 58
  in early code 155
uridine, *see* U *and codons/anticodons containing U*
uridine-5-oxyacetic acid (cmo^5U), base-pairing rules 16, 18–19
*UUC anticodon 20, 21
UUR codons 174
  yeast mitochondrial 104
UUU codon, identification 7
*UUU anticodon 20, 21

valine (Val) universal codons 10
  usage 60

wobble (hypothesis) 11, 12, 15
  four-way 25–6
  rules 11, 15

xm^5s^2U 19
  base-pairing rules 16
xo^5U, base-pairing rules 16, 18–19

yeast
  non-universal code 79–80, 81, 103–6, 107
    in mitochondria 103–6, 107
  phylogenetic relationships 79, 84
  $tRNA^{Arg}$ in mitochondria 26
  unassigned codons in mitochondria 59